BIOLOGY

FOR IB DIPLOMA COURSE PREPARATION

Marwa Bkerat

OXFORD
UNIVERSITY PRESS

Great Clarendon Street, Oxford, OX2 6DP, United Kingdom

Oxford University Press is a department of the University of Oxford. It furthers the University's objective of excellence in research, scholarship, and education by publishing worldwide. Oxford is a registered trade mark of Oxford University Press in the UK and in certain other countries

British Library Cataloguing in Publication Data
Data available

978-0-19-842350-8

10 9 8 7 6 5 4 3 2 1

Paper used in the production of this book is a natural, recyclable product made from wood grown in sustainable forests. The manufacturing process conforms to the environmental regulations of the country of origin.

Printed in India by Manipal Technologies Limited
Croydon CR0 4YY

Acknowledgements

The authors and publisher are grateful to those who have given permission to reproduce the following extract and adaptation of copyright material:

Francis Crick: from What Mad Pursuit: A Personal View of Scientific Discovery, copyright © 1988. Reprinted by permission of Basic Books., an imprint of Perseus Books, LLC, a subsidiary of Hachette Book Group, Inc.

We have made every effort to trace and contact all copyright holders before publication, but if notified of any errors or omissions, the publisher will be happy to rectify these at the earliest opportunity.

The publisher and the authors would like to thank the following for permission to use their photographs:

Cover: NG Images/Alamy Stock Photo

p20: Steve Gschmeissner/Science Photo Library; **p17:** Library of Congress Prints and Photographs Division; LC-DIG-ggbain-14544; **p41:** Photograph by A. Barrington Brown, Copyright Gonville and Caius College, Cambridge/Coloured by Science Photo Library; **p57:** Library of Congress Prints and Photographs Division; LC-USZ62-7261; **p66:** Bettmann/Getty Images; **p95:** Kateryna Kon/Shutterstock; **p96:** John Cairns/Oxford University Images/Science Photo Library; **p103 (L):** Dr_Microbe/iStockphoto; **p103 (L):** Tomasz Markowski/Shutterstock; **p104:** Jacopin/BSIP/Science Photo Library; **p106:** GL Archive/Alamy Stock Photo; **p112:** Magaiza/iStockphoto; **p121:** IFA Design, Plymouth, UK, Clive Goodyer, and Q2A Media.; **p105:** The Radboud University Nijmegen; www.vcbio.science.ru.nl; **p122:** Zeljko Radojko/Shutterstock; **p133:** Patila/Shutterstock; **p141:** Everett Historical/Shutterstock; **p146 (L):** Science Source/Science Photo Library; **p151:** Patrick Foto/Shutterstock; **p146 (R):** David Haykazyan/Shutterstock; **p149:** Cascade Creatives/Shutterstock. Artwork by Thomson Digital.

Contents

 Answers to questions in this book can be found at **www.oxfordsecondary.com/9780198423508**

The **Diploma Programme** (DP) is a two-year pre-university course for students in the 16–19 age group. In addition to offering a broad-based education and in-depth understanding of selected subjects, the course has a strong emphasis on developing intercultural competence, open-mindedness, communication skills and the ability to respect diverse points of view.

You may be reading this book during the first few months of the Diploma Programme or working through the book as a preparation for the course. You could be reading it to help you decide whether the Biology course is for you. Whatever your reasons, the book acts as a bridge from your earlier studies to DP Biology, to support your learning as you take on the challenge of the last stage of your school education.

Chapters 1 through to 6 of this book explain the biology that you need to understand at the beginning of a DP Biology course. You may already have met some of this science, but the book encourages you to begin to look at the concepts that underpin biology as a whole.

Chapter 7 of this book has advice on effective study habits and preparing for tests and examinations. Early preparation is vital even though now they may seem a long way off. The appendix lists information you will need throughout the DP Biology course, such as maths and ICT skills.

DP course structure

The DP covers six academic areas, including languages and literature, humanities and social sciences, mathematics, natural sciences and creative arts. Within each area, you can choose one or two disciplines that are of particular interest to you and that you intend to study further at the university level. Typically, three subjects are studied at higher level (HL, 240 teaching hours per subject) and the other three at standard level (SL, 150 hours).

In addition to the selected subjects, all DP students must complete three core elements of the course: theory of knowledge, extended essay, and creativity, activity, service.

Theory of knowledge (approximately 100 teaching hours) is focused on critical thinking and introduces you to the nature, structure and limitations of knowledge. An important goal of theory of knowledge is to establish links between different areas of shared and personal knowledge and make you more aware of how your own perspective might differ from those of others.

The **extended essay** is a structured and formally presented piece of writing of up to 4,000 words based on independent research in one of the approved DP disciplines. It is also possible to write an interdisciplinary extended essay that covers two DP subjects. The main purpose of the extended essay activity is to develop the high-level research and writing skills expected at university.

Creativity, activity, service involves a broad range of activities (typically 3–4 hours per week) that help you discover your own identity, adopt the ethical principles of the IB and become a responsible member of your community. These goals are achieved through participation in arts and creative thinking (creativity), physical exercises (activity) and voluntary work (service).

DP Biology syllabus
Basics

The DP Biology course itself is divided into four sections: Core syllabus, Additional Higher Level (AHL) material, and one of four possible Options together with an internal assessment (IA). There is also a group 4 project in which all science students in a school participate.

The Biology course is designed so that a student can study the entire course at standard level with no prior knowledge of biology. At higher level, however, some earlier study of the subject is advisable.

Biology standard level: Core + one Option at SL + IA + group 4 project

Biology higher level: Core + AHL + one Option at HL + IA + group 4 project

Biology topics

The *Biology Guide* is a document for teachers that lists the areas that students are to be taught as a series of *Topics*. On page viii there is a table showing the connections between this book and the distribution in the *Guide* of the various areas of biology between SL, HL and the Option topics.

Core topics

The six Core topics are taught and examined to the same standard at both levels. They are:

- Cell biology
- Molecular biology
- Genetics
- Ecology
- Evolution and biodiversity
- Human physiology

Many of these topics and sub-topics are covered in this book.

AHL topics

There are five additional higher-level topics, which you will only study if you have chosen to study biology at higher level:

- Nucleic acids (an extension of *Molecular biology*)
- Metabolism, cell respiration and photosynthesis (an extension of *Cell biology*)
- Plant biology (an extension of *Ecology*)
- Genetics and evolution (an extension of *Evolution and biodiversity*)
- Animal physiology (an extension of *Human physiology*)

Options

There are four Options, from which you will study one. Within each Option some topics correspond to Core areas and one or two additional areas are studied only by HL students. The Options are:

- Neurobiology and behaviour
- Biotechnology and bioinformatics imaging
- Ecology and conservation
- Human physiology

Internal assessment (IA)

Biology is an experimental science. The ability to plan and execute an experimental project is part of your assessment, which is where the **internal assessment** (IA) comes in. The internal assessment may include theoretical investigations and laboratory work. About 10 hours will be devoted to the IA, probably towards the end of the course. Your teacher will support you in carrying out the IA and you will be taught the required skills throughout the course. The IA accounts for 20% of your overall examination marks.

Group 4 project

Most students who study a group 4 subject undertake a collaborative project within – or possibly beyond – their school. Group 4 students work together on this project, for over about 10 hours. The project is not assessed formally. It emphasizes the relationships between sciences and how scientific knowledge affects other areas of knowledge. It can be experimental or theoretical. Be imaginative in your project, and perhaps combine with a different IB World School on another continent to study a project of mutual interest.

Aims of group 4

There are ten aims addressed by every group 4 subject. Each student should:

- be challenged and stimulated to appreciate science within a global context
- develop scientific knowledge and a set of scientific techniques
- apply and use the knowledge and techniques
- develop experimental and investigative scientific skills
- learn to create, analyse and evaluate scientific information
- learn to communicate effectively using modern communication skills
- realize the value of effective collaboration and communication in science
- have an awareness of the ethical implications of science
- appreciate the possibilities and limitations of science
- understand the relationships between scientific disciplines and between science and other areas of knowledge.

Key features of the DP Biology course

The following components are incorporated into the DP Biology course:

The **nature of science** (⊛) is the overarching theme in all IB science subjects, including biology. Throughout the course you will encounter many examples, activities and questions that go beyond the studied subject and demonstrate key principles of the scientific approach to exploring the natural world. For example, the development of cell theory was based on experimental evidence that had been accumulated by many generations of scientists and shared within the scientific community through collaboration and communication, bolstered by innovations in microscope technology (*1.1 Cell structure and function*).

Nature of science studies are not limited to the scientific method but cover many other aspects of science, from the uncertainty and limitations of scientific knowledge to the ethical and social implications of scientific research. Raising these issues will help you understand how science and scientists work in the 21st century.

Theory of knowledge (⊛) is another common feature of the DP Biology syllabus. In addition to the stand-alone theory of knowledge course taken by all DP students, much of the material in Biology topics can prompt wider discussions about the different ways of knowing used by scientists for interpreting experimental results.

Although theory of knowledge is not formally assessed in the DP Biology course, it facilitates the study of science, just as the study of science supports you in their theory of knowledge course.

International mindedness is one of the social aspects of science reflected in the IB mission statement, which emphasizes the importance of intercultural understanding and respect for creating a better and more peaceful world. International mindedness is actively promoted through all DP subjects by encouraging you to embrace diversity and adopt a global outlook.

Biology is an experimental science that provides you with numerous opportunities to develop a broad range of practical and theoretical skills.

Practical skills (⊛) are required for setting up experiments and collecting data. Typical laboratory works are described throughout this book and include measuring enzyme activity (*2.4 Proteins and enzymes*), quadrate sampling (*6.1 Ecosystems*), paper chromatography (*4.1 Photosynthesis*) and use of potometers to measure the rate of transpiration by plants (*4.2 Plant transport*).

Maths skills (⊛) are needed for processing experimental data and solving problems. In addition to elementary mathematics, the IB Biology syllabus requires the use of statistics, such as the calculation of standard deviation and error, and the use of charts and graphs. These maths skills are outlined in the appendix on page 160 of this book.

Approaches to learning (⊛) are a variety of skills, strategies and attitudes that you will be encouraged to develop throughout the course. The Diploma Programme recognizes five categories of such skills: communication, social, self-management, research and thinking. These skills are discussed in more detail in *7 Tips and advice on successful learning* on page 154 of this book.

Assessment overview

In addition to the internal assessment discussed earlier, the **external assessment** is carried out at the end of the DP Biology course. Both HL and SL students are expected to take three papers as part of their external assessment. You will usually take papers 1 and 2 at one sitting with paper 3 a day or two later. The question papers are as follows:*

Paper	SL	SL duration and marks	HL	HL duration and marks
1	30 multiple-choice questions on Core material	45 minutes; 30 marks; 20% of marks	40 multiple-choice questions on Core and AHL material	60 minutes; 40 marks; 20% of marks
2	Short and extended written answer questions on Core material; data-based question	75 minutes; 50 marks; 40% of marks	Short written answer and extended written answer questions on Core and AHL material; data-based question	135 minutes; 72 marks; 36% of marks
3	Section A: Data-based questions and questions based on experimental work Section B: Questions on your chosen Option	60 minutes; 35 marks; 20% of marks	Section A: Data-based questions and questions based on experimental work Section B: Questions on your chosen Option	75 minutes; 45 marks; 24% of marks

The internal and external assessment marks are combined to give your overall DP Biology grade, from 1 (lowest) to 7 (highest). The final score is calculated by combining grades for each of your six subjects. Theory of knowledge and extended essay components can collectively contribute up to three extra points to the overall Diploma score. Creativity, activity, service activities do not bring any points but must be authenticated for the award of the IB Diploma.

Using this book effectively

Throughout the book you will encounter separate text boxes to alert you to ideas and concepts. Here is an overview of these features and their icons:

Icon	Feature	Description of feature
WE	Worked example	A step-by-step explanation of how to approach and solve a biology problem.
Q	Question	A biology problem to solve independently. Answers to these questions can be found at **www.oxfordsecondary.com/9780198423508**
(key)	Key term	Defines an important scientific concept used in biology. It is important to be familiar with these terms to prepare you for the DP Biology course.
(DNA)	DP ready – Nature of science	Relates a topic in biology to the overarching principles of the scientific approach to exploring the natural world and the way discoveries are made.
(cap)	DP ready – Approaches to learning	Highlights the skills of an effective learner necessary for the DP.
(brain)	DP ready – Theory of knowledge	Features ideas or concepts in biology that prompt wider discussions about the different ways of knowing.
(infinity)	Internal link	Provides a reference to somewhere within this book with more information on a topic discussed in the text, given by the section number and the topic name. For example, *6.2 Energy and nutrient transfer: modes of nutrition* refers to the second section in Chapter 6 of this book and covers the ways organisms transfer energy and nutrients.
(link)	DP link	Provides a reference to a section of the DP Biology syllabus for further reading on a certain topic.
(calc)	Maths skills	Explains an important mathematical skill required for the DP Biology course.
(tool)	Practical skills	Relates the scientific theory to the practical aspects of biology you will encounter on the DP Biology course.

*Correct at the time of printing

Linking this book to the DP Biology syllabus

This textbook can be read linearly, but you might find it most useful to dip into specific sections to support different areas of your learning. For example, if you are at the start of your course, you might spend some time reading *7 Tips and advice on successful learning* to develop your study skills. Alternatively, if you are learning about Darwin's theory of evolution in class, read through the parts of *6 Ecology, evolution and classification* that explain the conceptual basis of evolution.

The following grid gives a comparison between the chapters of this book and all of the IB Diploma Programme Biology course topics.

DP Topic	Title	Sub-topics	Chapter in this book
1	Cell biology	1.1 Introduction to cells 1.2 Ultrastructure of cells 1.3 Membrane structure 1.4 Membrane transport 1.5 The origin of cells 1.6 Cell division	1
2	Molecular biology	2.1 Molecules to metabolism 2.2 Water 2.3 Carbohydrates and lipids 2.4 Proteins 2.5 Enzymes 2.6 Structure of DNA and RNA 2.7 DNA replication, transcription and translation 2.8 Cell respiration 2.9 Photosynthesis	2, 3, 4
3	Genetics	3.1 Genes 3.2 Chromosomes 3.3 Meiosis 3.4 Inheritance 3.5 Genetic modification and biotechnology	5
4	Ecology	4.1 Species, communities and ecosystems 4.2 Energy flow 4.3 Carbon cycling 4.4 Climate change	6
5	Evolution and biodiversity	5.1 Evidence for evolution 5.2 Natural selection 5.3 Classification of biodiversity 5.4 Cladistics	6
6	Human physiology	6.1 Digestion and absorption 6.2 The blood system 6.3 Defence against infectious disease 6.4 Gas exchange 6.5 Neurons and synapses 6.6 Hormones, homeostasis and reproduction	3
7	Nucleic acids (AHL)	7.1 DNA structure and replication 7.2 Transcription and gene expression 7.3 Translation	5 (very briefly)
8	Metabolism, cell respiration and photosynthesis (AHL)	8.1 Metabolism 8.2 Cell respiration 8.3 Photosynthesis	3,4 (very briefly)
9	Plant biology (AHL)	9.1 Transport in the xylem of plants 9.2 Transport in the phloem of plants 9.3 Growth in plants 9.4 Reproduction in plants	4 (briefly)
10	Genetics and evolution (AHL)	10.1 Meiosis 10.2 Inheritance 10.3 Gene pools and speciation	5, 6 (very briefly)
11	Animal physiology (AHL)	11.1 Antibody production and vaccination 11.2 Movement 11.3 The kidney and osmoregulation 11.4 Sexual reproduction	3

1 Cells

> " A cell has a history; its structure is inherited, it grows, divides, and, as in the embryo of higher animals, the products of division differentiate on complex lines. Living cells, moreover, transmit all that is involved in their complex heredity. "
>
> **Sir Frederick Gowland Hopkins, 'Some Aspects of Biochemistry',**
> ***The Irish Journal of Medical Science*** **(1932), 79, 346**

Chapter context

All living **organisms** are made of **cells**. Since the 17th century, tissues from different living organisms have been examined under **microscopes** and have shown that cells are the smallest unit of life. Some organisms are made of one cell, while others are made of many. Evolution has resulted in a great diversity between cells from the very simple **prokaryotic** cells to the most complex **eukaryotic** cells. Regardless of the differences between cells, there are many common features among them. All cells contain **genetic material, cytoplasm** and a **plasma membrane** that controls the composition of the cell. New cells come from pre-existing cells by **cell division**.

Learning objectives

In this chapter you will learn about:

→ the cell theory

→ the **basic structure** of cells

→ **transport** in cells

→ **cell division** in prokaryotes and eukaryotes.

Key terms introduced

→ Cells
→ Differentiation
→ Organelles
→ Prokaryotic cells and prokaryotes
→ Eukaryotic cells and eukaryotes
→ Magnification and resolution
→ Stem cells
→ Hydrophilic (polar) and hydrophobic (non-polar) substances
→ Integral and peripheral proteins
→ Aqueous, concentrated and dilute solutions
→ Hypotonic, hypertonic and isotonic solutions
→ Endosymbiotic theory
→ Interphase and DNA replication
→ Mitosis
→ Cytokinesis
→ Mutations and metastasis

1.1 Cell structure and function

Since the 17th century, microscopes have been used to examine tissues from different living organisms. This resulted in the development of the cell theory, which states that:

1. All living organisms are composed of *cells*.

2. Cells are the smallest unit of life.

3. Cells come from pre-existing cells and cannot be created from non-living material. Division of cells results in the formation of new cells.

Regardless of the differences between cells, all cells share some common features. All cells are surrounded by a *plasma membrane*, which separates the contents of the cell from its surroundings. All cells contain genetic material, which holds the information needed for the cell to carry out its activities. All cells contain *cytoplasm* where chemical reactions take place.

DP link

The structure and function of cells will be explained further in **1.1 Introduction to cells** in the IB Biology Diploma Programme.

Key term

Cells are the building blocks of life.

Trends and discrepancies

Most organisms conform to cell theory, some do not. The cell theory was based on the work of several scientists over many years where various trends among the cells of living organisms were discovered. Some discrepancies have been discovered but they were not enough to discard the cell theory. Many organisms consist of cells that are considered atypical. Examples of atypical cells include the striated muscle fibres which are larger than most animal cells and have many nuclei. Another example is giant algae (such as acetabularia) which are single-celled organisms with a much larger size than a normal cell.

Unicellular versus multicellular

Unicellular organisms, which are also known as single-celled organisms, are made up of a single cell. Examples of unicellular organisms include bacteria, amoeba, chlorella, paramecium and euglena. In unicellular organisms, the single cell is responsible for carrying out all the functions of life that are necessary for its survival. Table 1 indicates the seven functions of life that are necessary for the survival of any organism.

Table 1. The seven functions of life necessary for the survival of any organism

Function of life	Description
Metabolism	The chemical reactions that take place inside the cell
Response	The ability to react towards a stimulus
Homeostasis	Keeping the internal environment of the cell within limits
Growth	The increase in size
Reproduction	The production of offspring (sexual or asexual)
Excretion	The removal of waste products from the cell
Nutrition	Getting the material needed for growing and producing energy

Key term

The **plasma membrane (cell membrane)** surrounds the cell and separates the contents of the cell from its surroundings.

Cytoplasm is found within all cells, it is where the cellular chemical reactions take place.

Multicellular organisms, which are also known as multi-celled organisms, are made up of more than a single cell. Examples of multicellular organisms include plants and animals. The cells of multicellular organisms differentiate to make different tissues that perform specialized functions. For example, red blood cells are specialized to carry oxygen, whereas nerve cells are specialized to pass a nerve impulse.

Differentiation is the process by which a cell becomes more specialized. During the process of differentiation, some genes in the cell are "switched on". This means the gene starts to be used in the function of the cell, and we refer to the gene as being *expressed*. Other genes are switched off (or *unexpressed*). This results in cells that are more specialized and perform different functions. Differentiated cells form tissues, tissues form organs, organs form organ systems and organ systems form the multicellular organism.

Prokaryotes versus eukaryotes

Living organisms can be divided into two main groups based on the presence or absence of a nucleus and membrane-bound *organelles*: *prokaryotes* and *eukaryotes*.

Prokaryotic cells have a simple structure as they lack a nucleus and membrane-bound organelles. The genetic material (**DNA**) is not enclosed inside a **nucleus** but rather found in a region called the **nucleoid**. Prokaryotes include bacteria and archaea (ancient bacteria). *Escherichia coli* (*E. coli*) is an example of a bacterium. The structures found in most prokaryotic cells are described in figure 1 and table 2.

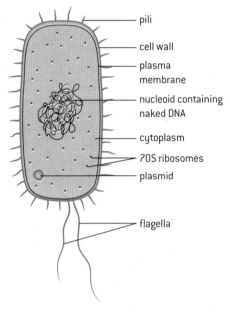

pili
cell wall
plasma membrane
nucleoid containing naked DNA
cytoplasm
70S ribosomes
plasmid
flagella

Figure 1. Prokaryotic cell structure (*E. coli*)

Table 2. The function of the main structures of prokaryotic cells

Structure	Description	Function
Cell wall	Made of peptidoglycan (a polysaccharide)	Maintains the shape of the cell and prevents the cell from bursting
Plasma membrane	A selectively permeable membrane	Controls the substances moving into and out of the cell
Cytoplasm	A gel-like substance enclosed within the cell	Contains enzymes to catalyse chemical reactions taking place inside the cell
Pili	Hair-like structures found on the surface	Help bacteria to adhere to each other for the transfer of DNA from one cell to another by a process called conjugation
Flagella (singular flagellum)	A whip-like structure	Helps bacteria move around
Ribosomes	70S type	Protein synthesis
Nucleoid	A region containing the naked DNA	Contains the DNA which holds the genetic information that controls the cell
Plasmid	A small ring of DNA	Helps bacteria adapt to unusual situations such as antibiotic resistance

Eukaryotic cells are more complex than prokaryotic cells as they contain a nucleus and membrane-bound organelles. The genetic material (DNA) is enclosed in a nucleus. Eukaryotes include plants, animals, fungi and protists.

Key term

Differentiation is the process by which a cell becomes more specialized. When a gene is switched on during this process, we say the gene is being expressed. When a gene is switched off, it is unexpressed.

DP link

The prokaryotic and eukaryotic cells will be explained further in **1.2 Ultrastructure of cells** in the IB Biology Diploma Programme.

Key term

Organelles are structures found inside cells that perform a specific function.

Key term

Prokaryotic cells are simple cells that lack a cell nucleus and membrane-bound organelles.

Prokaryotes, such as bacteria and archea, are single-celled organisms that do not contain a nucleus or any membrane-bound cell organelles.

Internal link

DNA will be explained in more detail in section **2.5 Nucleic acids** of this book.

Key term

Eukaryotic cells are complex cells that contain a cell nucleus and membrane-bound organelles.

Eukaryotes are single-celled or multicellular organisms whose cells contain a cell nucleus and membrane-bound cell organelles.

Eukaryotes may be unicellular or multicellular. An amoeba is an example of a unicellular eukaryote. Animals and plants are examples of multicellular eukaryotes.

The structures found in most eukaryotic cells in animals and plants are described in figure 2 and table 3, and figure 3 and table 4, respectively.

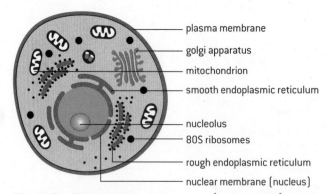

Figure 2. Eukaryotic cell structure (animal cell)

Table 3. The function of the main structures of eukaryotic cells (animal cells)

Structure	Description	Function
Ribosomes	Found either as 70S or 80S. Could be found free in the cytoplasm or attached to the rough endoplasmic reticulum	Protein synthesis
Smooth endoplasmic reticulum	No ribosomes on the surface	Lipid synthesis and transport
Rough endoplasmic reticulum	A network of tubules that extend from the nucleus to the rest of the cell	Protein synthesis and transport
Lysosome	Contains many enzymes	Digests waste structures within the cell such as dead organelles and foreign particles
Golgi apparatus	Consists of many flattened sacs stacked on top of each other. Has two sides, the cis side, which receives products from endoplasmic reticulum. The trans side, which is the side through which vesicles are released	Processing of proteins received from the rough endoplasmic reticulum. This includes packaging and modifying proteins to be used either inside the cell or excreted outside the cell
Mitochondrion	Contains its own ribosomes and DNA. It is made of two membranes: an outer membrane and an inner membrane that is folded inward to increase surface area	Production of ATP in aerobic respiration
Nucleus	It is surrounded by a porous double membrane	Contains the genetic material (DNA) which hold the genetic information that controls the cell
Nucleolus	Found inside the nucleus	Ribosomes synthesis

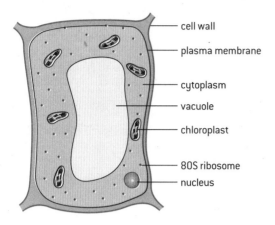

Figure 3. Eukaryotic cell structure (plant cell)

Table 4. Structures that are only found in plant cells

Structure	Description	Function
Cell wall	Made of cellulose (a polysaccharide)	Strengthens and supports the cell, maintains the shape of the cell and prevents the cell from bursting
Chloroplasts	Surrounded by two membranes. It contains its own ribosomes and DNA	Photosynthesis
Vacuoles	Storage organelles that come from the Golgi apparatus	Store water and food (cell sap)

Improved tools allow for new scientific discoveries
The electron microscope has a much greater magnification and higher resolution than the light microscope. It enables scientists to see the details of the organelles inside the cell.

Key term

Magnification is the size of image enlarged.

Resolution is the clarity of the view or image.

Table 5. Differences between prokaryotic cells and eukaryotic cells

Feature	Prokaryotic cells	Eukaryotic cells
Nucleus	No nucleus	Have nucleus
DNA	Found in the cytoplasm in a region named the nucleoid.	DNA found in the nucleus enclosed in a nuclear envelope.
	Circular DNA	Linear DNA
	Single strand	Double helix
Mitochondria	No mitochondria	Have mitochondria
Ribosomes	70S (smaller)	80S (larger)
Membrane-bound organelles	No membrane-bound organelles	Membrane-bound organelles such as Golgi apparatus and the endoplasmic reticulum
Plasmid	May have plasmid	No plasmid
Size	Small < 10 μm	Large > 10 μm
Complexity	Simple	Complex

Key term

Compare and contrast means to state the similarities and differences.

Compare means to state the similarities.

Distinguish means to state the differences.

Command terms and their meanings are available in the appendix.

Table 6. Differences between animal cells and plant cells

Feature	Animal cell	Plant cell
Cell wall	No cell wall	Have a cell wall (made of cellulose)
Chloroplasts	No chloroplasts	Have chloroplasts
Vacuoles	Do not usually contain any vacuoles and if present they are small or temporary	Have a large central vacuole
Shape	Rounded	Angular

Question

1 Distinguish between the genetic material present in prokaryotes and eukaryotes.

2 Compare and contrast the structure of animal cells with plant cells.

Maths skills: Standard form

You can use "standard form" to write very big or very small numbers in a more condensed form. When writing in "standard form" we use exponents of base 10.

Here are two examples:

1. The number 2 500 000 can be written as 2.5×10^6.

2. The number 0.0000543 can be written as 5.43×10^{-5}.

Follow these steps to express a number in standard form:

*   Write down the first few significant figures (numbers including and following the first digit that is not zero) that appear in the number as a number between 1 and 10 (the first example above would give 2.5, and the second example 5.43).

*   Write $\times 10$ after this number.

*   Count the number of places the decimal point would have to move from its original position to be between the first two significant figures of the number. This number becomes the exponent that you apply to 10.

*   If the decimal point is moved to the left, then the exponent is positive.

*   If the decimal point is moved to the right, then the exponent is negative.

So, for the examples above:

1. 2 500 000: In this case, the decimal point is at the end of the number (2 500 000.0). The decimal point would have to move six places to the left to be between digits 2 and 5, so it is therefore written as 2.5×10^6.

2. 0.0000543: The decimal point moves five places to the right, so it is therefore written as 5.43×10^{-5}.

Maths skills: Significant figures

The significant figures of a number are the digits of a number starting from the first non-zero digit. Significant figures are a good indication of the accuracy of a measurement. For example, if the number is 3.0 it indicates that the measurement was made accurate to the tenth (0.1) and if the number is 3.00 it indicates that the measurement was made accurate to the hundredth (0.01). Therefore, those numbers after the decimal point are considered significant.

For example:

- 2.3—has two significant figures
- 234—has three significant figures
- 5.0—has two significant figures
- 0.023—has two significant figures
- 1000—has one significant figure (it is unlikely the zeros are significant; they may have resulted from rounding off)
- 1000.0—has five significant figures (the number after the decimal point indicates that the measurement is accurate to the tenth).

When you get a number with many digits in a calculation, you often have to round it up or down to give the correct number of significant figures. The rules for rounding are:

- Round up if the number after your final significant figure is 5 or higher.
- Round down if the number after your final significant figure is 4 or lower.

In calculations, we give the answer to the same number of significant figures as the number used in the calculation with the least significant figures.

For example:

$34.322 \times 2.2 = 75.5084$

Since 2.2 has two significant figures, then the answer should be given as 76 (correct to two significant figures). The number following the second significant figure is 5, so we round up to 76.

Question

3 Write 75 330 000 in standard form.

4 Write 0.000074 in standard form.

5 How many significant figures does 0.045 have?

6 What is 4.564 to two significant figures?

Practical skills: Using a light microscope to observe onion cells

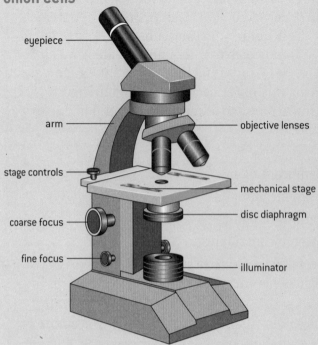

Figure 4. Compound light microscope

Onion cells are often chosen as an example of plant cells since they can be easily seen using a light microscope.

To prepare the slide to be observed, you need to cut the onion open and use forceps to peel a thin layer from the inside. Spread the thin layer on a microscope slide. Add a drop of iodine solution to stain the layer to make it easier to view its parts. Carefully place a cover slip over the layer and ensure no bubbles are trapped inside.

To view the slide, ensure that you move from one objective lens to another very carefully while adjusting the fine or coarse focus.

Maths skills: Calculating magnification and the actual size of cells

$$\text{Magnification} = \frac{\textit{image size}}{\textit{actual size}}$$

Image size is the measured size and can be found by measuring the micrograph size using a ruler.

Actual size is the real size of the cell which is either given or to be calculated.

Magnification will be given or must be calculated.

When solving questions, ensure that the same units are used for both the image and actual size.

Convert units when needed:

cm		mm		μm		nm
	x 10 → ÷ 10		x 1000 → ÷ 1000		x 1000 → ÷ 1000	

Worked example: Calculations involving magnification

1. Calculate the actual size of the dividing bacteria cell shown in figure 5.

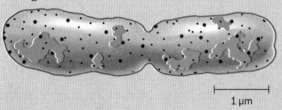

```
⊢————————⊣
     1 μm
```

Figure 5. A prokaryote undergoing binary fission

Solution

The first step is always to find the magnification. Since the magnification is not given, it must be calculated using the scale bar given.

$$\text{Magnification} = \frac{\textit{image size} \,(\textit{of the scale bar – measured})}{\textbf{actual size} \,(\textit{of the scale bar – indicated on the diagram})}$$

$$\text{Magnification} = \frac{1.5\,\text{cm}}{1\,\mu m}$$

Make sure the units are the same: 1.5 cm = 15 mm = 15 000 μm

$$\text{Magnification} = \frac{15\,000\,\mu m}{1\,\mu m}$$

Magnification = 15 000 ×

After finding the magnification, the actual size of the bacterium can be calculated by rearranging the formula used above, but this time using size of the amoeba.

$$\text{Actual size of the dividing bacterium} = \frac{\textit{image size}}{\textit{Magnification}}$$

$$\text{Actual size of the dividing bacterium} = \frac{7.5\,\text{cm}}{\times 15\,000}$$

Actual size of the dividing bacterium = 0.0005 cm
= 0.005 mm = 5 μm

Question

7 The length of a chloroplast is 50 μm and a drawing of it is 10 cm in length. Determine the magnification of the drawing.

8 Find the maximum length of the cheek cell in figure 6.

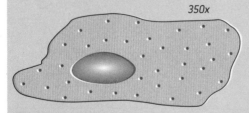
350x

Figure 6. Cheek cell

9 The magnification of an image is 2000 ×. The length of the drawn image is 30 mm. Calculate the actual length of the image.

DP link

Examples of diseases treated using stem cells will be discussed in detail in **1.1 Introduction to cells** in the IB Biology Diploma Programme.

Key term

Stem cells are undifferentiated cells that can divide and differentiate into different types of cells.

Plant stem cells are found in the **meristems** at the tips of roots and shoots.

Hematopoietic cells are stem cells which produce blood cells.

Stem cells

Stem cells are undifferentiated cells that can divide and differentiate into many different types of cell. Stem cells can be found in both plants and animals.

In plants, stem cells are found in *meristems* (the tips of roots and shoots).

In animals, stem cells can be found in different places including embryos, some adult cells such as *hematopoietic* cells (in bone marrow) and cord blood cells. Stem cells can be used for the treatment of many diseases such as leukemia, lymphoma and diabetes.

DP ready   **Nature of science**

Ethics and research

As research involving stem cells increases in importance, so does the need to understand and resolve the ethical issues such research raises. Many objections were raised against stem cell research, mainly embryonic stem cell research. Table 7 indicates the ethical considerations about the use of the three types of stem cells.

Table 7. The ethical considerations about the use of the three types of stem cells

Stem cells	Arguments for	Arguments against
Embryonic stem cells	• Easy to obtain • Pluripotent; can differentiate to many types of cells	• Involves the destruction of an embryo • May result in tumour development
Adult stem cells	• Does not involve the destruction of an embryo • Low chance of tumour development	• Limited differentiation • Sometimes not easily obtained
Cord blood stem cells	• Easy to obtain • Does not involve the destruction of an embryo • Can differentiate to many types of cells • Low chance of tumour development	• Must be obtained immediately after birth

Question

10 Discuss the ethical issues considered with the use of embryonic stem cells.

1.2 The cell membrane

The cell membrane is a complex structure that carries out a range of functions, such as acting as a barrier to keep cell contents inside the cell, and facilitating the transport of specific molecules into or out of the cell.

DP link

The membrane structure and transport will be studied in detail in **1.3 Membrane structure** and topic **1.4 Membrane transport** in the IB Biology Diploma Programme.

The phospholipid bilayer

The cell membrane is made up of a bilayer of phospholipid molecules (figure 7). Each phospholipid molecule is made of a head and a tail. The head is *hydrophilic* (attracted to water) and faces the outside of the cell whereas the tail is *hydrophobic* (repelled by water) and faces the inside of the membrane. This results in the phospholipid bilayer acting as a hydrophobic barrier; this means that *non-polar* (hydrophobic) substances can pass easily while *polar* (hydrophilic) substances cannot. Therefore, the phospholipid bilayer is a partially selective membrane. The structure of the phospholipid bilayer allows the cell membrane to change shape easily, which is required in endocytosis and exocytosis.

There are different types of membrane proteins that differ in structure, position and function. Some membrane proteins penetrate the phospholipid bilayer and are known as *integral proteins*. The integral proteins mostly act as pumps in active transport, or channels in facilitated diffusion, and therefore control the entry or exit of specific substances across the membrane. In contrast, there are membrane proteins that remain on the surface and are known as *peripheral proteins*. These proteins usually play a role in cell recognition which is involved in immune response.

Animal cell membranes contain cholesterol, which is a lipid component that is found in the hydrophobic region of the bilayer. Cholesterol has a role in decreasing the fluidity of the cell membrane and lowering its permeability to some molecules.

Key term

Hydrophilic substances are attracted to water and tend to be **polar**.

Hydrophobic substances are repelled by water and tend to be **non-polar**.

Internal link

Endocytosis and exocytosis will be discussed in **1.3 Cell transport**.

Key term

Integral proteins are membrane proteins that penetrate the phospholipid bilayer.

Peripheral protiens are membrane proteins that remain on the surface of the phopholipid bilayer.

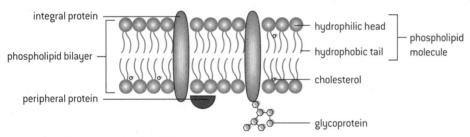

Figure 7. The phospholipid bilayer

DP ready   Nature of science

Scientists use models to represent concepts or ideas. There were various models that were developed to represent cell membrane structure. The phospholipid bilayer model shown in figure 7 was developed by Singer and Nicolson in 1972; this model is known as the fluid mosaic model or Singer–Nicolson model.

The Singer–Nicolson model was not the first model to describe the cell membrane structure. In 1935, Hugh Davson and James Danielli proposed a model for the cell membrane that was known as the lipo-protein sandwich model (figure 8). The model shows a phospholipid bilayer adjacent to layers of proteins on both sides of the membrane. When the cell membrane was viewed under the electron microscope, the electron micrograph showed two dark lines and a lighter band in between. Davson and Danielli proposed that the two dark lines are the protein layers on both sides of the membrane, and the light band is the phospholipid bilayer in the middle. The Davson–Danielli model was falsified after many experiments resulted in findings that did not fit with the model. For example, it was found that the membrane proteins are not fixed in a layer outside the phospholipid bilayer but rather are moving freely within the membrane.

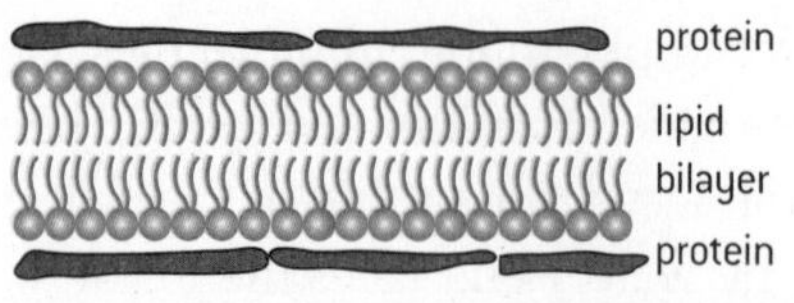

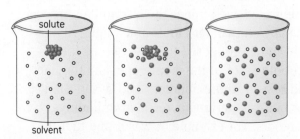

Figure 8. The Davson–Danielli lipoprotein sandwich model

The limitations noted with the Davson–Danielli model led to the proposal of the Singer–Nicolson model, or the fluid mosaic model. "Mosaic" refers to the different components that make up the membrane including proteins, carbohydrates and lipids. This model is currently considered the best model to describe the structure of the cell membrane.

DP ready   Theory of knowledge

Our understanding of the structure of the plasma membrane has changed over the years as new discoveries cause scientists to revise and update their theories. Why is it important to learn about theories that were later discredited?

1.3 Cell transport

Cell transport is the movement of particles across the cell membrane (the phospholipid bilayer). The phospholipid bilayer is a partially permeable membrane that allows some particles to pass through. Cell transport includes passive and active transport. Passive transport does not require energy and includes *simple diffusion*, *facilitated diffusion* and *osmosis*. Active transport requires energy and includes transport through pumps, *endocytosis* and *exocytosis*.

Particles move across cell membranes continuously. Such particles are usually found dissolved in a *solution*, mainly *aqueous solution*. This solution contains water as the solvent and various particles as solutes. If the amount of solute is high, the solution is described as a *concentrated solution*. If the amount of solute is little, the solution is described as a *dilute solution*.

Simple diffusion

Diffusion is the passive movement of particles from a region of high concentration to a region of low concentration until evenly distributed (figure 9). Diffusion does not require energy and takes place across a concentration gradient.

The concentration gradient is the difference in concentration between the two regions.

Key term

A **solution** is made up of a solute dissolved homogenously in a solvent.

An **aqueous solution** is a solution where the solvent is water.

A **concentrated solution** is a solution that has high amount of solute dissolved in the solvent.

A **dilute solution** is a solution that has a small amount of solute dissolved in the solvent.

solute

solvent

Figure 9. Diffusion

Which particles are transported by simple diffusion?

The movement of particles by diffusion across the cell membrane does not involve a channel and therefore, small hydrophobic (non-polar) molecules can easily pass through the phospholipid bilayer.

Examples of diffusion across the cell membrane (figure 10):

- the movement of O_2 from the blood capillaries to the body cells to carry out cellular respiration
- the movement of CO_2 (which is a by-product of cellular respiration) from the body cells to the blood capillaries to be transported and removed out of the body.

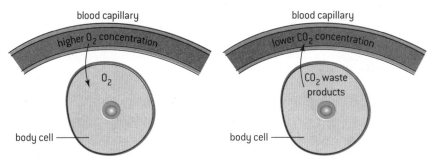

Figure 10. Oxygen and carbon dioxide diffusion

Osmosis

Osmosis is the passive movement of water molecules, across a partially permeable membrane, from a region of lower solute concentration (*hypotonic*) to a region of higher solute concentration (*hypertonic*) (figure 11). Water molecules are small enough to pass through the partially permeable cell membrane. Osmosis can also maintain a balanced or *isotonic solution* where the solute concentration level inside the cell is the same as the solute concentration level outside the cell.

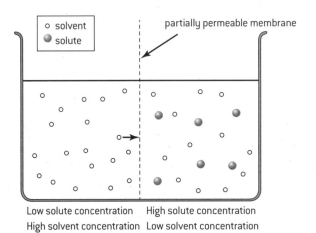

Figure 11. Osmosis

Key term

A **hypotonic solution** is a solution that has lower solute concentration.

A **hypertonic solution** is a solution that has higher solute concentration.

An **isotonic solution** is a solution that has the same solute concentration.

Osmosis in plant cells

The cell membrane of the plant cell acts as a partially permeable membrane. The cell sap inside the vacuole is a highly concentrated solution. Plant cells are surrounded by a rigid cell wall.

When a plant cell is placed in a hypotonic (lower solute concentration) solution, water molecules move from the solution into the cell by osmosis. This causes the cell to swell but the cell wall prevents it from bursting. The plant cell becomes **turgid**.

When a plant cell is placed in a hypertonic (higher solute concentration) solution, water molecules move out of the cell by osmosis. This causes the cell to shrink and pull away from the cell wall. The plant cell becomes **flaccid**. When the cytoplasm is pulled away from the cell wall, the cell becomes **plasmolysed** (figure 12).

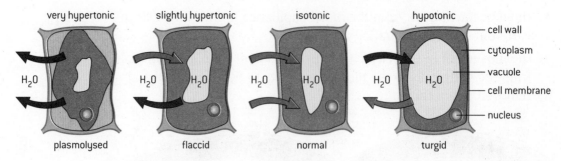

Figure 12. Osmosis in plant cells

When an animal cell is placed in a hypotonic (lower solute concentration) solution, water molecules move from the solution into the cell by osmosis. This causes the cell to swell up and explode as there is no cell wall to prevent the cell from bursting.

When an animal cell is placed in a hypertonic (higher solute concentration) solution, water molecules move out of the cell by osmosis. This causes the cell to shrink and become **shrivelled** (figure 13).

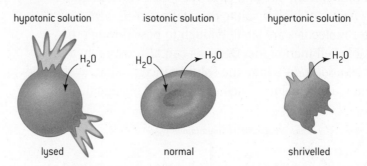

Figure 13. Osmosis in animal cells

Practical skills: Osmosis in potato cells

You can observe the effect of osmosis in potato cells by placing potato strips in different concentrations of saline (salt) solution. The concentrations must range from pure water to highly concentrated saline solution. Ensure that you control some variables such as the volume of the solution, size of potato strips, time soaking the strips in the saline solution and temperature. You need to identify your dependent variable; what are you measuring? You can measure the mass of the potato strips before and after placing them in the saline solution. Collect observations about the texture and colour of the potato strips—these observations are your qualitative data. Record the mass of the potato strips before and after placing them for a period of time in the different saline concentrations—these measurements are your quantitative data.

Question

11 Potato strips were soaked in a highly concentrated saline solution. How do you think this will affect the mass of the potato strips? Explain your answer.

Facilitated diffusion

Facilitated diffusion is the passive movement of particles from a region of high concentration to a region of low concentration via a channel protein (figure 14). Molecules that are large and polar cannot directly pass through the phospholipid bilayer and therefore require a channel protein that is embedded in the membrane to transport them.

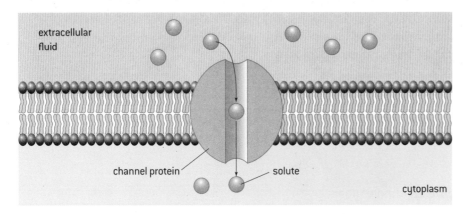

Figure 14. Facilitated diffusion

An example of facilitated diffusion across a cell membrane is the diffusion of potassium ions across the axon of a neuron through potassium channels. Neurons, also known as nerve cells, are the primary components of the nervous system.

Question

12 Compare and contrast between simple diffusion and facilitated diffusion.

Active transport

Active transport is the movement of substances through a membrane from a region of low concentration to a region of high concentration. This process occurs against a concentration gradient and therefore requires energy and protein pumps that are embedded in the cell membrane.

Each protein pump is specific and can only transport specific substances. The pump is provided with energy from ATP. ATP molecules that attach to the pump cause a conformational change in the shape of the pump, and molecules that are large and polar may be transported by active transport.

Internal link

ATP was introduced in **1.1 Cell structure and function**.

An example of active transport is the sodium–potassium pump which is embedded in the cell membrane of the axon in neurons (figure 15).

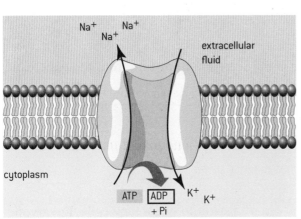

Figure 15. The sodium–potassium pump

DP link

The movement of potassium and sodium ions across the axon of the neuron will be discussed in **6.5 Neurons and synapses** in the IB Biology Diploma Programme.

The pump transports three sodium ions to the outside of the axon and two potassium ions into the axon.

Endocytosis

Endocytosis is the process by which large molecules enter the cell. This process requires energy.

The process begins when the cell membrane is pulled inwards due its fluidity. A vesicle pinches off into the cell membrane, carrying the material to be taken into the cell. The vesicle enters the cell and releases its contents (figure 16).

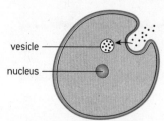

Figure 16. Endocytosis

Internal link

Phagocytosis will be discussed further in **3.4 Body defence**.

An example of endocytosis is phagocytosis. This is when foreign organisms such as bacteria are engulfed by macrophages (a type of white blood cell), that have a role in immune response. The bacterium is engulfed by endocytosis, and then it is moved towards the lysosome where it is digested by enzymes.

Exocytosis

Exocytosis is the process by which large molecules are released out of the cell. This process requires energy.

The process begins when the proteins synthesized in the rough endoplasmic reticulum are released in vesicles that are transported to the cis side of the Golgi apparatus for further modification. The vesicles carrying the proteins bud off the Golgi apparatus and are moved towards the cell membrane. The vesicle fuses with the cell membrane due to its fluidity, thus releasing the materials as it carries them out of the cell (figure 17).

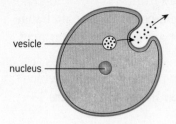

Figure 17. Exocytosis

An example of exocytosis is the secretion of insulin from the pancreatic cells into the blood stream.

Question

13 Explain why liver cells contain high amounts of mitochondria.

1.4 Origin of cells

Prokaryotic cells divide by binary fission to make new cells. Eukaryotic cells divide by mitosis to produce genetically identical body cells, and by meiosis to produce sex cells including sperm and eggs.

DP link

The origin of cells will be explained further in **1.5 The origin of cells** in the IB Biology Diploma Programme.

Evidence from experiments conducted by various scientists in the past have shown that cells come from the division of pre-existing cells.

Louis Pasteur (figure 18), the French chemist and microbiologist, designed an experiment to test if microbes could generate spontaneously inside chicken broth. Pasteur was able to disprove the theory of spontaneous generation developed by Aristotle by conducting his famous swan-neck flask experiment.

Internal link

Mitosis will be discussed in **1.5 Cell division**.
Meiosis will be discussed in **5.2 Reproduction and meiosis**.

Figure 18. Louis Pasteur (1822–1895)

In his experiment, Pasteur boiled chicken broth in two set-ups of swan-neck flasks to ensure that all microorganisms were killed. In the first set-up, the neck of the flask was broken off and the broth inside became contaminated (experiment 1, figure 19). In the second set-up, the curved neck of the flask prevented air from contaminating the broth inside, and so the broth inside remained sterile (experiment 2, figure 19). Pasteur concluded that cells come from pre-existing cells and do not spontaneously generate. This was a great discovery with regard to microbes and the origin of cells.

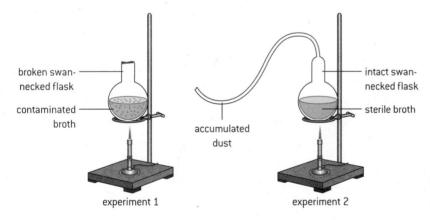

Figure 19. Pasteur's experiments

Pasteur's experiments provided evidence that cells come from pre-existing cells, but how were the first cells formed? Many scientists believe that the first cells may have come from non-living material. In 1950, biochemists Stanley Miller and Harold Urey were able to produce simple organic compounds from inorganic compounds when simulating the conditions of the Earth.

Key term

Endosymbiotic theory is an evolutionary theory that describes the endosymbiotic relationship between organisms and how it resulted in the evolution of eukaryotic cells from prokaryotic cells.

DP link

Cell division will be explained further in **1.6 Cell division** in the IB Biology Diploma Programme.

The origin of eukaryotic cells

Scientists use the *endosymbiotic theory* to describe how eukaryotic cells evolved from prokaryotic cells through a symbiotic relationship between organisms. It is believed that a large host cell engulfed a bacterium through endocytosis which resulted in a symbiotic relationship where both cells benefited from each other for survival. Scientists believe that mitochondria evolved from energy-producing bacteria that were engulfed by larger bacteria to supply the cell with the energy needed, while chloroplasts evolved from photosynthetic bacteria that were engulfed by larger bacteria to supply the cell with food.

Evidence that supports the development of mitochondria and chloroplasts by the process described in the endosymbiotic theory includes:

- Both organelles have their own DNA
- Both organelles have 70S ribosomes similar to prokaryotes
- Both organelles have the same size and shape of bacteria
- Both organelles have a double membrane.

1.5 Cell division

Cells reach a certain size and then divide. As cells grow, the surface area to volume ratio decreases. The surface area of the cell controls the rate of material exchange, while the volume of the cell controls the rate of resource consumption, and waste and heat production. When the cell grows, the volume increases much faster than the surface area (so the ratio decreases). As a result, material cannot be exchanged fast enough to provide enough resources for the cell to consume, and to get rid of the waste and heat produced in the cell. Therefore, the cell stops growing and divides.

The cell cycle

The cell cycle describes the behaviour of cells as they grow and divide. The cell cycle involves three main stages, *interphase*, *mitosis* and *cytokinesis* (figure 20).

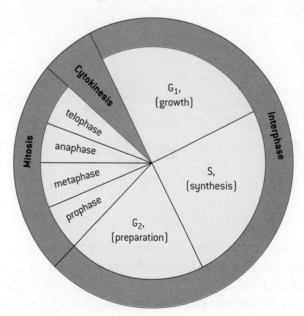

Figure 20. The cell cycle

Stage 1—Interphase

Interphase is the longest phase of the cell cycle. During this phase, the nucleus and cytoplasm pass through many processes such as *DNA replication*, protein synthesis and an increase in the number of organelles. It is divided into three phases: G1, S and G2. The main activities in each phase include:

- G1 phase: The cell grows in size, protein synthesis takes place in the cytoplasm and organelles increase in number.
- S phase: DNA replication takes place in the nucleus.
- G2 phase: The cell grows in size and prepares for mitosis.

Stage 2—Mitosis

Mitosis is the division of the nucleus to form two genetically identical *daughter nuclei*. Mitosis is divided into four main phases: prophase, metaphase, anaphase and telophase (figure 21).

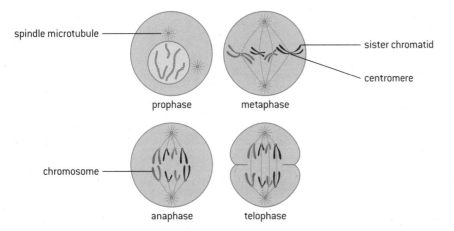

Figure 21. The four phases of mitosis

1 **Prophase**
- The spindle microtubules (a component of the cytoplasm) start growing and extend from each pole to the equator.
- *Sister chromatids* condense, thicken, shorten and become visible.
- The nuclear membrane starts breaking down.

2 **Metaphase**
- The sister chromatids move to the equator and line up separately.
- The spindle microtubules from each pole attach to each *centromere* on opposite sides.
- The spindle microtubules are fully developed.

3 **Anaphase**
- The spindle microtubules contract to pull the sister chromatids apart splitting the centromeres.
- This splits the sister chromatids into chromosomes.
- Each identical chromosome is pulled to opposite poles.

4 **Telophase**
- The spindle microtubules break down.
- The chromosomes decondense and are no longer individually visible.
- The nuclear membrane reforms.

 Key term

Interphase is the longest phase of the cell cycle.

DNA replication is the production of two identical copies of DNA.

 Internal link

DNA replication will be studied in detail in **2.5 Nucleic acids**.

 Key term

Sister chromatids are the identical copies formed by the replication (duplication) of a chromosome. They are joined together by what is known as a **centromere**.

When sister chromatids separate, we refer to them as chromosomes.

Key term

Mitosis refers to the division of the nucleus into two genetically identical daughter nuclei.

Key term

Cytokinesis is the division of the cytoplasm to split the cell into two daughter cells. Cytokinesis takes place in meiosis and mitosis.

The cell then divides by cytokinesis to form two daughter cells with identical genetic nuclei.

Mitosis occurs in processes such as growth, embryonic development, tissue repair and asexual reproduction.

Stage 3—Cytokinesis

Cytokinesis is the last stage of the cell cycle during which the cytoplasm divides to develop two identical daughter cells. In plant cells, the cell wall forms a *plate* between the dividing cells. In animal cells, the cell pinches in the middle of the two dividing cells forming a *cleavage furrow* (figure 22).

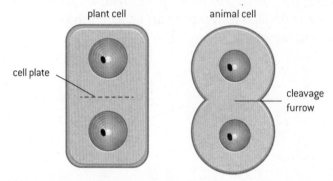

Figure 22. Cytokinesis in plant and animal cells

Maths skills: Calculating the mitotic index

The mitotic index is a tool that is used to predict the response of cells to chemotherapy. Chemotherapy is used as a treatment for cancer cells. A high mitotic index indicates that there are many cells dividing, which is the case with cancer cells.

The mitotic index can be calculated using the formula:

$$\frac{\text{number of cells in mitosis counted in a micrograph}}{\text{total number of cells in a micrograph}} \times 100\%$$

Worked example: Calculating the mitotic index

2. Calculate the mitotic index of the micrograph below (figure 23):

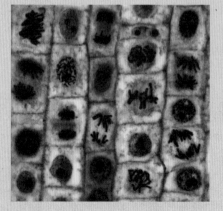

Figure 23. Micrograph of cells undergoing mitosis

Solution

1. Count the number of cells undergoing mitosis = 13
2. Count the total number of cells = 25
3. Use the formula:

$$\frac{\text{number of cells in mitosis counted in a micrograph}}{\text{total number of cells in a micrograph}} \times 100\%$$

Mitotic index $= \dfrac{13}{25} \times 100 = 52\%$

Question

14 Distinguish between cytokinesis in plant cells and animal cells.

Cell division in prokaryotes

Most prokaryotic cells divide by binary fission. Binary fission is an asexual method that involves the splitting of the parent cell to produce two genetically identical cells (figure 24).

Binary fission involves the following steps:

1. DNA replication takes place in which the DNA strand and the plasmid are duplicated.
2. Each DNA copy separates and moves to an opposite direction.
3. The cell grows and elongates.
4. Cytokinesis: the cell membrane pinches in the middle to divide the cell into two identical daughter cells.

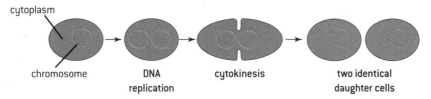

cytoplasm

chromosome · DNA replication · cytokinesis · two identical daughter cells

Figure 24. Binary fission

Tumour formation

Cell division is a controlled process in the body. Tumours form because of uncontrolled cell division. There are two types of tumours:

- **Benign tumours** which do not invade any body tissue. These tumours are usually inactive and harmless.
- **Malignant tumours** which are detached and carried in the bloodstream to other parts of the body. This will result in a secondary tumour or *metastasis*. These tumours are known as **cancer** because they are dangerous and life-threatening.

What causes tumour formation?

The cell cycle is controlled by specific genes. If these genes mutate, the cell cycle is no longer controlled, and the cell will divide in an abnormal way. *Mutation* occurs because of several factors including smoking, high exposure to radiation such as X-rays and *ultraviolet (UV) light*, some viruses, and carcinogens such as benzene and asbestos.

Key term

Mutations are random changes to the base sequence of genes.

Metastasis is the movement of cells from a primary tumour that are carried in the bloodstream and form secondary tumours in other parts of the body.

UV light or **ultraviolet light** is electromagnetic radiation of wavelengths shorter than visible light. It is found between the violet light and X-rays on the electromagnetic spectrum.

Chapter summary

In this chapter, you have learned about the cell theory, the basic structure of cells, transport in cells and cell division in prokaryotes and eukaryotes. Make sure that you have a working knowledge of the following concepts and definitions:

- ☐ Cells are the smallest unit of life of which all living organisms are made.
- ☐ Unicellular organisms are made up of a single cell and carry out the seven functions of life.
- ☐ Multicellular organisms are made up of many cells that differentiate to perform specialized functions.
- ☐ Prokaryotic cells have a simple structure as they lack a nucleus and membrane-bound organelles, whereas eukaryotic cells have a more complex structure as they have a nucleus and membrane-bound organelles.
- ☐ Animal cells and plant cells are examples of eukaryotic cells but they have some differences.
- ☐ Stem cells are undifferentiated cells that can divide and differentiate into many different types of cells and can be used in the treatment of many diseases.
- ☐ Cell transport is the movement of particles across the cell membrane.
- ☐ The cell membrane is made up of a bilayer of phospholipid molecules, which acts as a hydrophobic barrier that controls the material going in and out of the cell.
- ☐ Cell transport includes passive and active transport.
- ☐ Passive transport does not require energy and includes simple diffusion, facilitated diffusion and osmosis.
- ☐ Active transport requires energy and includes transport through pumps, endocytosis and exocytosis.
- ☐ The surface area to volume ratio limits the growth of the cell and therefore the cell divides when it reaches a certain size.
- ☐ The cell cycle describes the behaviour of cells as they grow and divide, and involves three main stages: interphase, mitosis and cytokinesis.
- ☐ Prokaryotes divide by binary fission.
- ☐ Tumours form because of uncontrolled cell division.

Additional questions

1. Evaluate the cell theory.
2. Distinguish between the light microscope and the electron microscope.
3. Compare and contrast between prokaryotes and eukaryotes.
4. Calculate the magnification of a sperm cell that has a tail measuring 40 µm in length which is drawn as 4 cm long.
5. State the name of the process by which bacteria pass their genetic material from one cell to another.
6. List the factors that determine the ease by which a molecule crosses the plasma membrane.
7. Deduce what will happen if a skin cell is placed in a concentrated saline solution for a long period of time.
8. Compare and contrast between passive and active transport.
9. Explain why *E. coli* does not burst when it is placed in water.
10. Explain how the surface area to volume ratio limits the growth of cells.
11. State where DNA replication takes place.
12. Describe the stages involved in mitosis.
13. State which processes involve mitosis.
14. Explain tumour formation.

2 Biological molecules

> " Almost all aspects of life are engineered at the molecular level, and without understanding molecules we can only have a very sketchy understanding of life itself. "

Francis Crick, 1988

Chapter context

Biological molecules include all the molecules within living organisms that are responsible for the processes of life and the interactions within the various systems of a cell. Within living organisms, there are many **essential elements** that are required for life; they are the building blocks of the biological molecules that make up the living matter. **Carbon, hydrogen, oxygen** and **nitrogen** make up the basic structure of the biological molecules in living organisms. Water makes up about 60% of the human body and is considered the medium of life. **Carbon compounds** are the basis of life and they are commonly made up of carbon, hydrogen, oxygen and nitrogen atoms.

Learning objectives

In this chapter you will learn about:

→ the essential **elements** for life

→ the **structure of water** and its function in living organisms

→ the structure and function of **carbohydrates, lipids, proteins, enzymes** and **nucleic acids**.

Key terms introduced

→ Atoms and elements

→ Carbon (organic) compounds and hydrocarbons

→ Covalent and ionic bonds

→ Electronegativity

→ Hydrogen bonds

→ Mono-, di-, poly- and sacchar- prefixes

→ Condensation and hydrolysis

→ Hydrogenation of oils

→ Amino acids and polypeptides

→ Peptide bonds, disulfide bridges and denatured proteins

→ Substrates and active sites

→ Optimum temperature

→ Gene

→ Codon

2.1 Essential elements for life

There are around 25 elements which are known to be essential for life.

Essential elements are needed in specific amounts; some are needed in large amounts while others are needed in small amounts. Hydrogen, oxygen, carbon and nitrogen are considered the most frequently occurring elements in living organisms as they are needed for the basic structure of *carbohydrates*, *lipids*, *proteins* and *nucleic acids*. Other elements are required in small amounts, though they have a major role in keeping the body working effectively. Such elements include iron, calcium, phosphorus and sodium. Table 1 indicates some of the essential elements and the role of each inside human bodies.

Key term

An **essential element** is any element that is required for life and its absence causes abnormal development or functioning.

Table 1. Some essential elements and their function

Element	Function
Sulfur	Needed for the synthesis of amino acids
Calcium	Acts as a co-factor in some enzymes. It is also a component of bones
Phosphorus	Needed for the following: • formation of the nucleotides in DNA molecules • formation of phospholipids • formation of ATP
Iron	Needed for the synthesis of hemoglobin, which is a protein that carries oxygen in the blood
Sodium	Needed for osmotic balance. It is also needed in sending nerve impulses

Carbon

Many scientists describe life as "carbon-based", which means that if carbon was not present, life on Earth would not have been possible. This is because carbon is required for the formation of most of the molecules present in living organisms.

Carbon is a relatively small *atom*. It has six neutrons and six protons inside its nucleus. It has two electrons in its inner shell and four electrons in its outermost shell (figure 1).

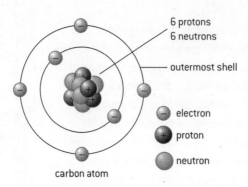

6 protons
6 neutrons

outermost shell

− electron

+ proton

neutron

carbon atom

Figure 1. Carbon atom

Key term

Atoms are the building blocks of all matter on Earth. There are many kinds that differ in mass, size and internal structure, and they can combine in different ways.

An **element** is a substance made of atoms of the same kind.

Carbon can form four *covalent bonds* with itself or other elements such as **hydrogen**, **oxygen**, **nitrogen**, **sulfur** and others. The simplest carbon molecule is methane (CH_4), in which carbon binds to four hydrogen atoms, as shown in figure 2.

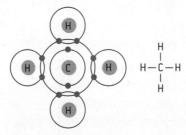

Figure 2. Two ways of representing methane

Key term

Carbon compounds (organic compounds) are compounds that are found in living organisms and contain carbon. Carbon compounds include carbohydrates, lipids, proteins and nucleic acids.

Hydrocarbons are a large family of organic molecules that are composed of hydrogen atoms bonded to a chain of carbon atoms. Methane is an example of a hydrocarbon.

Carbon can form single, double or triple bonds. It can form long branched and unbranched chains, and even rings. It can bond with itself to form extremely strong and stable compounds. The ability of carbon to form different types of bonds enables it to form many different *carbon compounds*, that range from the simplest *hydrocarbons* to the most complex carbon compounds that can be found in living organisms.

Question

1 Explain why carbon dioxide is not considered an organic compound.

2 Explain why carbon can form four covalent bonds with itself or other elements.

2.2 Water

Water is of major importance to all living organisms. Most of the body weight of living organisms comes from water. In humans, it constitutes 60% of an adult's body weight.

The structure of water

The water molecule (H_2O) is composed of one oxygen atom linked by a single bond to two oxygen atoms. Each hydrogen atom shares a pair of electrons with the oxygen atom. Therefore, the bond between each hydrogen atom and the oxygen atom is a *covalent bond*. The two hydrogen atoms are bound to the oxygen atom at an angle of 104.5° to each other. This gives the water molecule a "bent" structure (figure 3).

Key term

A **covalent bond** is a chemical bond that involves the sharing of electron pairs between atoms.

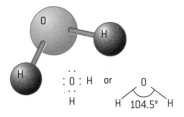

Figure 3. The water molecule

The oxygen atom has a greater *electronegativity* than a hydrogen atom which means that the oxygen atom will pull the shared electrons more towards it. This will cause the oxygen atom to have a partial negative charge ($\delta-$) and the hydrogen atoms to have a partial positive charge ($\delta+$). Therefore, the bond between the hydrogen and oxygen atoms is a polar covalent bond, which makes the water molecule polar (figure 4).

Key term

Electronegativity is the ability of an atom to attract an electron pair, shared with another atom in a chemical bond, to itself.

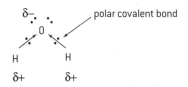

Figure 4. Polar covalent bond in water molecule

Due to the opposite charges at the two ends, the positive hydrogen end of one molecule is attracted to the negative oxygen end of another molecule. This bond is called a *hydrogen bond*, which is approximately ten times weaker than the average covalent bond. Each water molecule binds to four other water molecules by hydrogen bonding (figure 5). The hydrogen bond between water molecules gives water its unique properties.

Key term

A **hydrogen bond** is a bond between the positive hydrogen end of one molecule and the negative oxygen end of another molecule. They are weaker than covalent bonds.

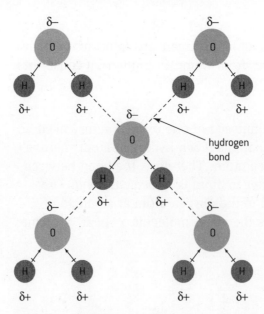

Figure 5. Hydrogen bonding between water molecules

The properties of water

Water is a transparent, tasteless and odourless liquid at room temperature. It is the most abundant substance on Earth and can be found as a solid, liquid or gas. It has unique properties due to the presence of the hydrogen bonds between its molecules. Its melting point is 0°C and its boiling point is 100°C. When water is in its solid form, ice, it is less dense than liquid water, which explains why glaciers float in water. It is described as the "universal solvent" because of its ability to dissolve many substances. It has a high specific heat capacity, which means that it would require high amounts of energy to increase its temperature. It also has a high latent heat of vaporization, which means that it needs to absorb a high amount of heat to evaporate. Water is cohesive, which means that water molecules tend to bond to itself by forming hydrogen bonds. Water is adhesive, which means that water molecules are attracted to other types of molecules due to its polarity.

> **DP ready   Nature of science**
>
> **Using theories to explain natural phenomena**
> Scientists come up with theories to explain why a natural phenomenon may occur. For example, scientists in the early 20th century proposed the theory that hydrogen bonds form between water molecules. Though hydrogen bonds are not visible directly, their existence explains the unique properties of water detailed previously. We can assume a theory is correct if we cannot falsify it, and if there is experimental evidence to support it.

The functions of water

Water has several important functions in the body:

* Cell life: Water makes up most of the cell and therefore it is a vital nutrient to the life of every cell.

- Transport medium: Some substances are hydrophilic (water loving) and so they dissolve in water. For example, oxygen and nutrients such as glucose are transported to the cells whereas waste products such as urea are eliminated from the body through urine.

- Metabolic reactions: Many chemical reactions in the body require water. For instance, water is involved in *hydrolysis* reactions in which macromolecules such as proteins, lipids and carbohydrates are broken down into simpler molecules. Such reactions cannot take place without water.

- Temperature regulation: Water regulates our internal body temperature by sweating. When sweat evaporates, the body cools down and constant body temperature is maintained.

Question

3   State which bond is responsible for the unique properties of water molecules and describe these properties.

2.3 Carbohydrates and lipids

Carbohydrates

Carbohydrates are the body's main source of energy. They contain carbon, hydrogen and oxygen, and they can be represented by the formula $(CH_2O)_n$, where n indicates the number of carbon atoms in the molecule. The ratio of carbon to hydrogen to oxygen is always 1:2:1. Carbohydrates can be classified into three main groups: *monosaccharides*, *disaccharides* and *polysaccharides*.

Monosaccharides are the simplest forms of carbohydrates. The number of carbon atoms in monosaccharides range between three and six in each molecule. The names of most monosaccharides end with the suffix -ose. Monosaccharides can be divided based on the number of carbon atoms into trioses (three carbon atoms), pentoses (five carbon atoms) and hexoses (six carbon atoms). Glucose, galactose and fructose are examples of hexoses, while ribose and deoxyribose are examples of pentoses (figure 6). Table 2 indicates some examples of monosaccharides and their function.

> **Key term**
>
> The prefixes **mono-, di-** and **poly-** are often used in biological terms.
>
> - **mono-** means "one"
> - **di-** means "two"
> - **poly-** means "many"
>
> Another commonly used term is **sacchar-** which means "sugar".

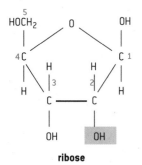
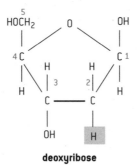

Figure 6. Ribose and deoxyribose (pentose sugars)

Table 2. Examples of monosaccharides and their function

Monosaccharide	Function
Glucose	An important source of energy. It is used in aerobic respiration to help make ATP
Galactose	Nutritive sweetener in foods
Fructose	Fruit sugar
Ribose	It is part of ribonucleic acid (RNA)
Deoxyribose	It is part of deoxyribonucleic acid (DNA)

Glucose, galactose and fructose have the same chemical formula $(C_6H_{12}O_6)$ but they differ slightly in their structures (figure 7).

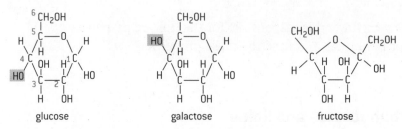

Figure 7. Glucose, galactose and fructose (monosaccharides)

Internal link

Aerobic respiration and using glucose to make ATP will be discussed in **3.3 Breathing and respiration**.

Internal link

Absorption of nutrients including monosaccharides will be discussed in **3.1 The digestive system**.

Monosaccharides, like other nutrients, are absorbed in the small intestine. During digestion, all carbohydrates must be broken down into monosaccharides in order to be absorbed. Thus, monosaccharides are the building blocks of all carbohydrates.

Disaccharides form when two monosaccharides are linked together via a covalent bond. This reaction is called *condensation* because it involves the release of water. The most common disaccharides include maltose, lactose and sucrose. Lactose consists of the monosaccharides, glucose and galactose. Sucrose consists of the monosaccharides glucose and fructose. Table 3 indicates the most common disaccharides and their function.

Key term

Condensation is the formation of macromolecules from simple molecules which involves the release of water.

Hydrolysis is the breakdown of macromolecules into simpler molecules which involves the use of water.

Table 3. Examples of disaccharides and their function

Disaccharide	Function
Maltose	Also called malt sugar. It is found in barley.
Lactose	The sugar found naturally in milk.
Sucrose	Also called table sugar. It is the transport sugar in plants.

Polysaccharides form when many monosaccharides are linked by covalent bonds to form a long chain, which can be branched or unbranched. Polysaccharides are very large molecules and they are usually insoluble in water. Starch, glycogen and cellulose are the most common polysaccharides (table 4).

DP link

Condensation and hydrolysis reactions and the structure of polysaccharides will be studied in **2.3 Carbohydrates and lipids** and **2.4 Proteins** in the IB Biology Diploma Programme.

Table 4. Examples of polysaccharides and their function

Polysaccharide	Function
Starch	Storage of carbohydrates in plants
Glycogen	Storage of carbohydrates in animals
Cellulose	Main component in plant cell walls

Question

4 Describe the three types of carbohydrate and give one example for each.

Lipids

Lipids include a wide range of compounds that are hydrophobic (water fearing) and thus insoluble in water. Lipids include fats, oils, waxes, phospholipids and steroids. The different types of lipids have different structures and perform different functions in organisms.

- Lipids can be used as long-term energy storage in the form of fats.
- Lipids provide heat insulation for animals that have a large fat layer under their skin.
- Lipids allow buoyancy as they are less dense than water and therefore enable animals to float on water.
- Lipids form many hormones in the body such as testosterone.
- Phospholipids are an important constituent of the plasma membrane.

Fats are also known as triglycerides, they consist of two main parts: a glycerol molecule and three fatty acids. The glycerol molecule is a small organic molecule that is made of three carbon atoms, five hydrogen atoms, and three hydroxyl (–OH) groups. Fatty acids have a long hydrocarbon chain attached to an acidic carboxyl (–COOH) group. Fatty acids may contain up to 36 carbon atoms. In a fat molecule (a triglyceride), a fatty acid is attached to the glycerol molecule at each of the three hydroxyl (–OH) groups via a covalent bond (figure 8).

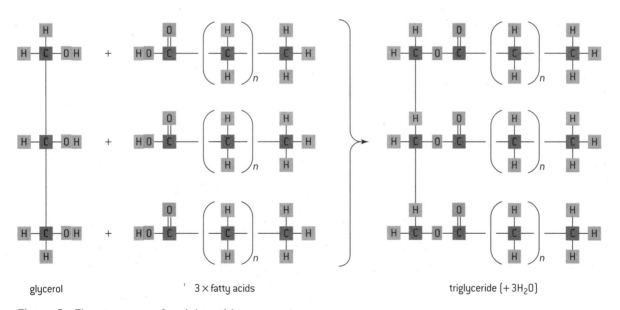

glycerol   3 × fatty acids   triglyceride (+ 3H₂O)

Figure 8. The structure of a triglyceride

Saturated and unsaturated fatty acids

Fatty acids are the main components of most lipids. They consist of a covalently bonded carbon chain, and at one end they have a methyl group (–CH₃), while at the other end they have a carboxyl (–COOH) group (figure 9).

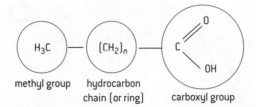

methyl group | hydrocarbon chain (or ring) | carboxyl group

Figure 9. The basic structure of fatty acids

Fatty acid chains may differ in length (number of carbon atoms), as well as in the presence and location of double bonds. The C–C bond holds a great deal of energy. Therefore, the longer the fatty acid chain, the more energy it holds. Fatty acids are categorized into two main groups: saturated fatty acids and unsaturated fatty acids (figure 10).

- Saturated fatty acids have only single bonds between neighbouring carbon atoms in the hydrocarbon chain. This means that they are saturated with hydrogen. Saturated fats are solid at room temperature and are usually found in foods such as meat and butter.

- Unsaturated fatty acids have one or more double bonds between neighbouring carbon atoms in the hydrocarbon chain. This means that they have fewer hydrogen atoms and therefore can gain hydrogen atoms (or be *hydrogenated*). If the unsaturated fatty acid has just one double bond, it is monounsaturated, while if it has multiple double bonds, it is polyunsaturated. Most unsaturated fats are liquid at room temperature and are found in oils such as vegetable and fish oil.

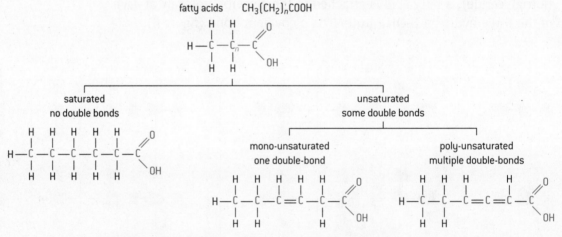

Figure 10. Saturated and unsaturated fatty acids

There are two configurations for the unsaturated fatty acids based on the location of hydrogen atoms in the double bond: the *cis* configuration and the *trans* configuration (figure 11).

- Cis unsaturated fatty acids: Where the hydrogen atoms are on the same side of the double bond. A cis configuration will cause a twist at the double bond, a feature that results in the cis unsaturated fats being liquid at room temperature. Cis unsaturated fats are naturally found in food such as olive oil.

- Trans unsaturated fatty acids: Where the hydrogen atoms are on the opposite side of the double bond. A trans configuration will cause the fatty acid to be straight and therefore the molecules pack tightly against one another. This will cause the trans unsaturated fatty acids to be solid at room temperature.

Trans unsaturated fatty acids are artificially made in the food industry by hydrogenating the unsaturated fats. During the *hydrogenation process,* the naturally occurring cis configuration is converted to the trans configuration. Trans unsaturated fats include the hydrogenated oils found in some processed foods like margarine.

> **Key term**
>
> **Hydrogenation of oils** occurs when hydrogen molecules are used to saturate unsaturated fats in the presence of a catalyst.

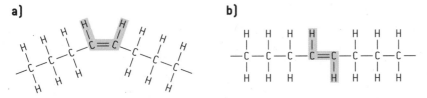

Figure 11. a) Cis configuration and **b)** trans configuration of fatty acids

> **Question**
>
> 5   Describe the different structures of fatty acids.

Fatty acids and human health

Cis unsaturated fatty acids help to improve blood cholesterol levels, whereas artificially made trans unsaturated fatty acids and saturated fatty acids contribute to *plaque* formation in the arteries, which increases the risk of *coronary heart disease* (CHD).

> **Key term**
>
> **Coronary heart disease (CHD)** is a disease in which a waxy substance called plaque builds up inside the coronary arteries, which supply oxygen to the heart muscle.

> **DP ready   Theory of knowledge**
>
> There are many claims about the harms and benefits of fats on human health. There is a positive correlation between the intake of saturated fatty acids and the rates of CHD. However, a correlation is not a proof of cause. For instance, Masai people of Kenya who depend on high saturated fat diets, have one of the lowest incidences of CHD. Scientists claim that other factors may be correlated to the intake of saturated fats such as fibre intake, lifestyle and genetic factors. Additionally, consuming of trans saturated fatty acids is positively correlated to the rates of CHD—in patients who died from CHD, a high proportion of fats found in the diseased arteries were trans fats.

Body mass index (BMI)

Body mass index (BMI) is a measure of body fat based on weight in relation to height. It is a screening tool that can be used to identify possible weight problems in adult men and women. BMI is not a direct measure of body fat, but it correlates to the amount of fat in the body. However, BMI does not account for factors such as age, gender and muscle mass. BMI is interpreted by the categories shown in table 5.

Table 5. BMI categories

BMI	Category
Below 18.5	Underweight
18.5–24.9	Normal
25.0–29.9	Overweight
30.0 and above	Obese

Maths skills: How to determine your body mass index (BMI)

BMI can be determined in two ways:

1. Calculate your BMI using the formula below:

$$BMI = \frac{body\ mass\ (kg)}{(height)^2\ (m^2)}$$

The unit for BMI is kg m^{-2}.

2. Use the BMI nomogram (figure 12) as follows:
   - Find your height in the left-hand column.
   - Find your weight in the right-hand column.
   - Use a ruler to draw a line between the two values.
   - The value of intersection is your BMI.

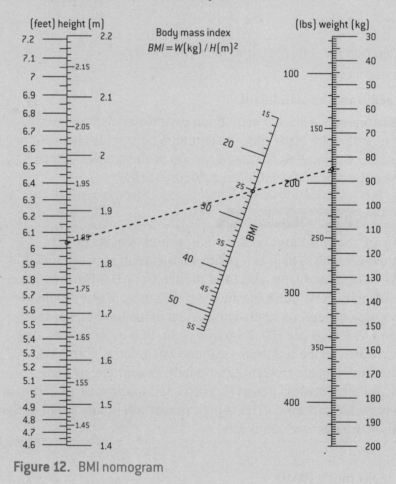

Figure 12. BMI nomogram

Worked example: Calculating BMI

1. Calculate the BMI for a person whose weight is 70 kg and height is 1.75 m. To which BMI category does he belong to?

Solution

$$BMI = \frac{70\,kg}{(1.75)^2\ m^2} = 22.9\ kg\ m^{-2}$$

This person's BMI is normal.

2. Using the BMI nomogram in figure 12, find the BMI for a person whose weight is 103 kg and height 1.95 m. To which BMI category does he belong to?

Solution

Use the BMI nomogram as follows:

1. Identify the height value of 1.95 m in the left-hand column.
2. Identify the weight value of 103 kg in the right-hand column.
3. Use a ruler to draw a line between the two values.
4. The value of intersection is 27 kg m^{-2}. This is the BMI.

This person is overweight.

Question

6 Calculate the BMI for a person whose weight is 65 kg and height is 1.80 m. To which BMI category do they belong to?

7 An adult man whose weight is 90 kg and height is 1.7 m, is considered obese. Explain why, using the BMI nomogram in figure 12.

Phospholipids

Phospholipids are the major component of the cell membrane. They are composed of two fatty acid chains attached to a glycerol, which is attached to a phosphate group (figure 13). The phospholipid molecule therefore has both hydrophobic and hydrophilic regions.

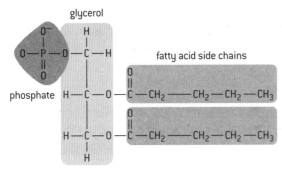

Figure 13. Phospholipid molecule

Waxes

Wax is a hydrophobic lipid which prevents water from sticking to it. Wax forms a protective coat on the surface of the leaves of many plants and on the body of some animals, such as the feathers of some birds.

Steroids

Steroids do not resemble other lipids in their structures due to the presence of a ring structure. However, they are considered lipids because they are hydrophobic. Cholesterol is an example of a steroid. Cholesterol is synthesized in the liver and is a key component in the formation of bile and the plasma membranes of animal cells.

Question

8 Describe the various functions of lipids in organisms.

DP link

The role of wax in conserving water in plants is studied in **9.1 Transport in the xylem of plants (AHL)** in the IB Biology Diploma Programme.

DP link

The role of cholesterol in reducing membrane fluidity and permeability to some solutes is studied in **1.3 Membrane structure** in the IB Biology Diploma Programme.

2.4 Proteins and enzymes

Proteins

Proteins are large carbon compounds composed of long chains of *amino acids* called *polypeptides*. Proteins perform important functions in living things and they are essential for life. That is why we need to regularly consume food that contains proteins. Such food includes beans, meat, milk and fish. There are different types of proteins in an organism or even in a cell. They are more diverse in their structures and function than any other macromolecule in living organisms.

Functions of proteins

Proteins perform different functions in living organisms. Some examples of the functions of proteins include:

- Proteins maintain the structure of the body—for example, collagen strengthens bones and skin.

- Some proteins act as *enzymes* and speed up metabolic reactions—for example, amylase is an enzyme that breaks down starch to simple molecules.

- Some proteins act as hormones to trigger a specific response in the body to control body functions—for example, insulin is a hormone that regulates blood sugar in the body.

- Proteins have a major role in immune response—for example, antibodies assist in destroying foreign particles in the body such as bacteria and viruses.

- Proteins facilitate the transportation of certain molecules in the body—for example, hemoglobin transports oxygen throughout the body.

Structure of proteins

The building blocks of all proteins are amino acids. Proteins vary in their structure from a very simple amino acid chain to a very complex subunit. Figure 14 shows the basic structure of an amino acid. Each amino acid is made of three main parts: an amino group (–NH$_2$), a carboxyl group (–COOH) and an "R-group". The identity of this R-group varies between different amino acids—this is what distinguishes one amino acid from another.

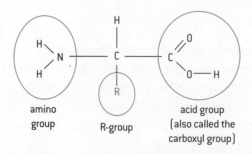

Figure 14. The basic structure of an amino acid

Amino acids are linked together by a *peptide bond* to form polypeptide chains. A protein may consist of a single polypeptide or more than one polypeptide linked together. The long polypeptide chains that form proteins fold and contort to create three-dimensional shapes. There are four main types (or structures) used to describe the folding and overall structure of polypeptides: primary, secondary, tertiary and quaternary.

The primary structure describes the unique sequence of amino acids linked together by peptide bonds, making a polypeptide chain.

The secondary structure describes the way in which the primary structure is folded, giving a distinct shape to specific regions of the chain. There are two main shapes, created by the formation of hydrogen bonds linking the adjacent amino acids in the folded areas of the polypeptide chain:

- The α-helix is a spiral shape created by the coiling of a polypeptide chain
- The ß-pleated sheet forms when the polypeptide chain folds in parallel or antiparallel layers.

Tertiary structure refers to the overall three-dimensional shape of an entire protein molecule. This folding results in the formation of several types of bonds that maintain the structure of the protein. Bonds may include hydrogen bonds, *ionic bonds* and *disulfide bridges*. Most enzymes have an overall rounded or globular shape created by the tertiary structure.

Quaternary structure refers to the interactions between multiple polypeptide chains, in proteins made from more than one polypeptide chain. In these complex proteins, each polypeptide chain is known as a subunit. These proteins may also include non-protein subunits such as heme in hemoglobin.

When proteins are exposed to extreme conditions such as heat, or strong acids or bases, they become *denatured*. This means that proteins will lose their secondary and tertiary structures; when these extreme conditions are implemented, the bonds within the protein structure break and this results in the altering of the shape of the protein. When the protein denatures, it returns to its primary structure.

 Key term

The **peptide bond** is a strong covalent bond that links amino acids in a polypeptide chain.

An **ionic bond** is a strong bond that forms between positive and negative ions.

Disulfide bridges form between two amino acids that contain sulfide.

A **denatured** protein is one that has lost its secondary or tertiary structure as a result of damage from exposure to extreme conditons.

Practical skills: Drawing molecular diagrams

You should be able to draw the diagrams of some of the molecules discussed by knowing some basics related to the molecule structure (such as the number of carbon atoms, linear or ring, etc.) and the names and structures of the chemical groups (functional groups) that may be attached to it. Some of these functional groups that are attached to molecules as part of their structure are shown in table 6 on page 36.

Table 6. Names and structures of some chemical groups found in biological molecules

Name of chemical group	Structure	Simplified notation
Hydroxyl group	—O—H	—OH
Amine group (also named amino group)	—N, with H above and H below	$-NH_2$
Carboxyl group (also named acid group)	—C, with =O above and O—H below	—COOH
Methyl group	—C— with H above, H below, and H to the right	$-CH_3$

Consider the following points when drawing molecules:

- Represent each atom in a molecule using the symbol of the element. For example, represent a carbon atom with C and a hydrogen atom with H.
- A line represents a single covalent bond while two lines represent a double bond.
- Attach the appropriate chemical groups based on what you know about the basic structure of a molecule. For example, an amino acid must include an amine group on one end. So, attach ($-NH_2$) on one end.

Question

9   List and describe the functions of proteins.

Enzymes

Enzymes are biological catalysts that speed up chemical reactions within cells. They are essential for life as they are needed for important functions in the body such as digestion. Some enzymes may help break down large molecules into smaller ones, while others help join smaller molecules to make larger ones. The molecule that binds to the enzyme is called the *substrate*. All enzymes have an *active site* to which a specific substrate binds.

Properties of enzymes

All enzymes share the following properties:

- They are all proteins folded into complex shapes.
- They are specific, which means that each enzyme can only catalyse one reaction because only a substrate molecule with a specific shape can bind to the active site.
- Like all proteins, they are affected by pH and temperature.
- They are not used up in reactions and therefore can be reused.

Key term

The **substrate** is the molecule that fits the active site.

The **active site** is the area on an enzyme to which the specific substrate binds.

Internal link

Digestive enzymes and their roles will be discussed in **3.1 The digestive system**.

How do enzymes work?

One of the theories that explains the interaction between the enzyme and substrate is the lock and key theory, where the enzyme is the lock and the substrate is the key (figure 15). This is because the shape of the active site is complementary to the shape of the substrate. Once a specific substrate fits the active site of the enzyme, an enzyme–substrate complex forms. The bonds within the substrate are weakened which results in the formation of a new product.

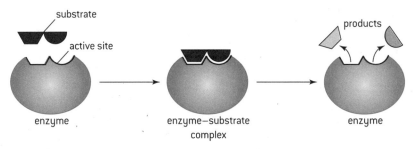

Figure 15. The lock and key theory

Enzymes and temperature

All enzymes are affected by temperature (figure 16). As the temperature increases, the enzyme activity and the rate of reaction increase. At the *optimum temperature*, the maximum rate of reaction is reached. Beyond the optimum temperature, the enzyme begins to lose its shape, including the shape of the active site. Therefore, the enzyme will no longer fit the substrate. When an enzyme loses its shape and is unable to function it is said to be denatured. Denaturing can occur when the enzyme is exposed to high temperatures or extreme pH.

Enzymes and pH

There is no one optimum pH for all enzymes. The optimum pH for an enzyme depends on where it normally works. For example, enzymes that work in the stomach have a very low optimum pH of about 2, while enzymes that work in the small intestine have an approximately neutral pH of about 7.5 (figure 17).

Most enzymes in the human body have an optimum pH near to 7.5. At the optimum pH, the enzyme works most efficiently, and the maximum rate of reaction is reached. Above or below the optimum pH, the rate of reaction and enzyme activity decrease. If the conditions are too acidic or too basic an enzyme will lose its shape and denature (figure 18).

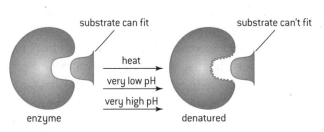

Figure 18. Enzyme denatured

DP link

The induced-fit model which was suggested by Daniel Koshland in 1958 is another model that is a more precise version of the lock and key theory. It explains why some enzymes can bind to many different substrates. This model is studied in
2.5 Enzymes in the IB Biology Diploma Programme.

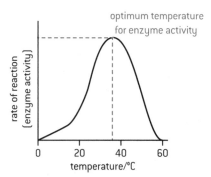

Figure 16. Effect of temperature on enzyme activity

Key term

The **optimum temperature** is the temperature at which the maximum rate of reaction is reached.

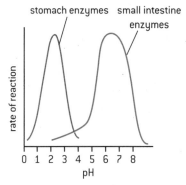

Figure 17. Effect of pH on enzyme activity

Enzymes and substrate concentration

The change in substrate concentration affects the rate of reaction. As the substrate concentration increases, the rate of reaction increases as well. This is due to the increase in collisions between the substrate and the enzyme. However, at a certain point when all enzyme molecules are occupied, the rate of reaction reaches its maximum and any further increase in substrate concentration will no longer affect the rate of reaction (figure 19).

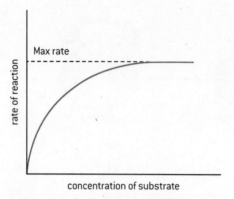

Figure 19. Effect of substrate concentration on the rate of an enzyme-catalysed reaction

Enzymes and industry

Being able to manipulate and control the temperature, pH and substrate concentration can affect the activity of the enzyme and consequently the rate of reaction. This makes it possible for us to control enzyme-catalysed reactions in industry to reach the maximum rate of reaction and increase the amount of products produced.

Lactase is an enzyme that can be obtained from fungi and is used in industry to produce lactose-free milk. Lactose is the sugar of milk, and it cannot be absorbed by the small intestine because it is a disaccharide. Lactase is an enzyme that breaks down lactose into glucose and galactose which are easily absorbed by the small intestine.

$$\text{Lactose} \xrightarrow{\text{Lactase}} \text{glucose} + \text{galactose}$$
$$\text{(disaccharide)} \qquad\qquad \text{(monosaccharides)}$$

People with lactose intolerance lack the enzyme lactase and so cannot break down lactose. If milk or milk-based products enter the digestive system, they are not digested. Instead, bacteria in the large intestine feed directly on lactose resulting in abdominal pain, diarrhea, cramps and excessive gas.

The production of lactose-free milk allows people who are lactose intolerant to consume milk. In addition, galactose and glucose taste sweeter than lactose. One of the ways to produce lactose-free milk is to fix the enzyme to small immobilized beads, where milk is poured in at the top. This results in the enzyme breaking down lactose in milk into glucose and galactose.

The development of some techniques may be of greater benefit to some particular human populations more than others. For instance, the development of lactose-free milk is of larger benefit to African and Asian people than European and American people, as lactose intolerance is more prevalent in Africans and Asians. Techniques developed in one part of the world may be more applicable in another.

Practical skills: Investigate the effect of pH or temperature on the activity of enzymes

You can plan an investigation to test the effect of factors including pH and temperature on the activity of enzymes, using a variety of different enzyme sources to carry out the investigation. For example, catalase is found in several sources including cow liver. In your DP Biology class, you may investigate the effect of temperature on the activity of catalase, with the knowledge that catalase is an enzyme that breaks down hydrogen peroxide into water and oxygen.

In this case, the independent variable would be temperature. The dependent variable is the variable that you measure. You can measure the volume of oxygen gas produced to learn about the relationship between catalase activity and temperature. Other variables we need to consider are the control variables, which are kept constant to ensure a fair test. These include the volume and concentration of hydrogen peroxide used and the source of catalase.

Question

10 Outline how enzymes are specific.

2.5 Nucleic acids

Nucleic acids are macromolecules that hold the genetic information which provide instructions for the formation of proteins essential for life. This information is passed from parents to offspring. There are two types of nucleic acid: deoxyribonucleic acid (DNA) and ribonucleic acid (RNA).

Nucleotides

Nucleic acids are composed of *nucleotides* that are linked together. Each nucleotide consists of three main parts (figure 20):

1. A phosphate group
2. A pentose sugar (ribose in RNA, deoxyribose in DNA)
3. A nitrogenous base, which could be adenine, guanine, thymine, or cytosine (uracil replaces thymine in RNA).

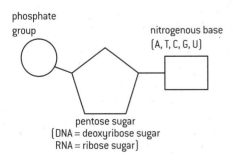

Figure 20. Nucleotide structure

RNA structure

RNA is a single-stranded nucleic acid that consists of ribose sugar. It has the same nitrogen bases as DNA with the exception of thymine which is replaced by uracil. It exists in several forms, and each has a unique function in the body. For example, messenger RNA (mRNA) and transfer RNA (tRNA) play important roles in the synthesis of proteins. Ribosomal RNA (rRNA) is a component of the ribosome which is the site for protein synthesis.

DNA structure

The DNA molecule is made of two strands that wind around each other like a twisted ladder, called a *double helix* structure. Each strand consists of nucleotides which are linked together via covalent bonds. The sugar and phosphate groups form the backbone of the double helix, while the nitrogen bases are found in between the two strands attached to each sugar. The two strands are held together by a hydrogen bond that links the bases together, where adenine pairs with thymine and cytosine pairs with guanine. This is referred to as complementary base pairing. Adenine forms two hydrogen bonds with thymine whereas guanine forms three hydrogen bonds with cytosine. Although the hydrogen bonds between the two strands are weaker than the covalent bonds, they are strong enough to keep the double helix in shape (figure 21).

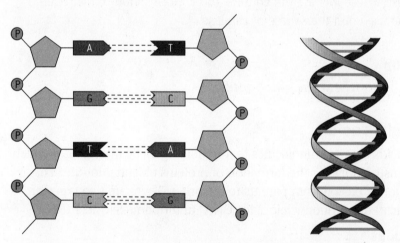

Figure 21. Left: Structure of DNA; right: DNA double helix

DP link

The structure of DNA and RNA will be explained further in **2.6 Structure of DNA and RNA** in the IB Biology Diploma Programme

Table 7. Similarities and differences between DNA and RNA

Feature	DNA	RNA
Similarities	<ul><li>both are nucleic acids</li><li>both consist of nucleotides which contain a pentose sugar, a nitrogenous base and a phosphate group</li></ul>	
Differences	Double-stranded	Single-stranded
	The pentose sugar is deoxyribose	The pentose sugar is ribose
	The nitrogen bases include: adenine, guanine, cytosine, thymine	The nitrogen bases include: adenine, guanine, cytosine, uracil

Using models as representation of the real world

In 1962, James Watson and Francis Crick (along with physicist Maurice Wilkins) were awarded the Nobel Prize in Medicine for their discovery of the structure of DNA as a double helix. Their model helped to explain how DNA replicates and how genetic information is passed to offspring. This was one of the most significant scientific discoveries of the 20th century. However, Crick and Watson's discovery was based on previous research conducted by several scientists including Rosalind Franklin, who studied DNA using X-ray crystallography.

Figure 22. Watson and Crick's DNA model

Question

11 Draw a labelled diagram to show four DNA nucleotides, each with a different base, linked together in two strands.

DNA replication

DNA replication is the process by which the DNA is copied inside a cell as part of its preparation for cell division (figure 23). This process takes place in the cell nucleus during the *interphase* stage of the cell cycle. The process takes place as follows:

1. The DNA double helix is separated into two individual strands of DNA by the enzyme helicase. Helicase does this by unwinding the double helix and breaking the hydrogen bonds that hold the DNA strands together.

2. The new strand of DNA is created by the enzyme, DNA polymerase. DNA polymerase uses one of the original DNA strands as a template and links nucleotides together, forming a new strand. Complementary base pairing, in which adenine is bonded to thymine and guanine is bonded to cytosine, is maintained.

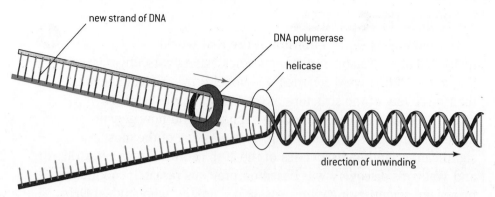

new strand of DNA

DNA polymerase

helicase

direction of unwinding

Figure 23. DNA replication

DNA replication is semi-conservative, meaning that each strand in the DNA double helix acts as a template for the formation of a new complementary strand (figure 24).

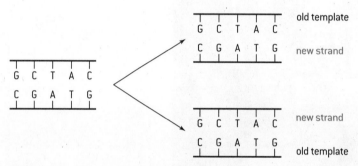

old template

G C T A C

C G A T G new strand

G C T A C

C G A T G

new strand

G C T A C

C G A T G old template

Figure 24. DNA replication is semi-conservative

Protein synthesis

Protein synthesis is one of the most vital biological processes, in which organisms build proteins which are essential for life. The process is initiated in the nucleus but then it is moved to the cell cytoplasm. DNA is the molecule that holds the genetic information that gives the instructions to build proteins. Since DNA cannot leave the cell, it sends a messenger outside to code for the formation of polypeptides. This messenger is called *messenger RNA* or mRNA. Therefore, protein synthesis involves two processes:

1. Transcription = mRNA formation
2. Translation = Polypeptide formation.

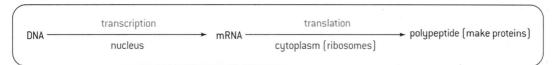

Figure 25. Transcription and translation in protein synthesis

DNA transcription

Transcription is the copying of a DNA sequence to make mRNA, using the catalyst RNA polymerase (figure 26). It involves the following stages:

1. RNA polymerase uncoils a section (*gene*) of the DNA double helix.

2. RNA polymerase links free RNA nucleotides to form an RNA strand (using a DNA strand as a template). This is done through complementary base pairing, however, in the RNA chain, the base thymine is replaced by uracil.

3. The mRNA strand then elongates and separates from the DNA template.

4. The DNA strands then reform a double helix.

5. The mRNA leaves the nucleus and moves out through the pores on the nuclear membrane.

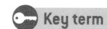 **Key term**

A **gene** is a section of a DNA molecule that serves as the basic unit of heredity.

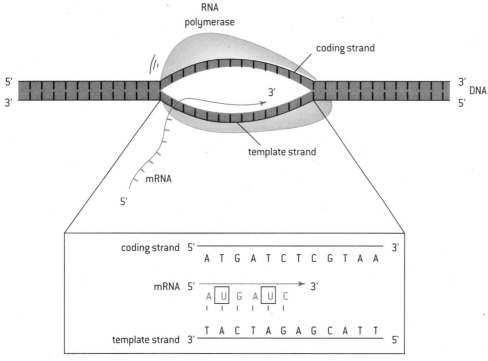

Figure 26. DNA transcription: The template strand is complementary to mRNA. The coding strand is a strand of DNA, so its bases are A, T, C and G; the mRNA reads the template strand so the base pair sequence is similar to the DNA coding strand but the mRNA has U instead of T

12 Compare the processes of DNA replication and transcription.

The genetic code

Each mRNA molecule carries a genetic code, which is defined by the sequence of nucleotides on the mRNA molecule. These nucleotides are arranged into triplets of base pairs knowns as *codons*. Here are some features of codons:

Key term

A **codon** is a sequence of three nucleotides that corresponds to a specific amino acid.

- There are 64 different codons.
- Each codon codes for a particular amino acid. There are 20 amino acids.
- There is a start codon at the start of the gene.
- There is a stop codon at the end of the gene.
- The code on mRNA determines the sequence of amino acids and the polypeptide formed.
- Codons will code for the same amino acid in all living things.
- Different codons can code for the same amino acid.

DNA translation

Translation is the synthesis of polypeptides on *ribosomes* in the cytoplasm (free ribosomes or ribosomes attached to the rough endoplasmic reticulum) (figure 27). The genetic code of the mRNA determines the amino acid sequence of the polypeptide. As discussed previously, the genetic code is made of codons. Each codon consists of three bases that correspond to a specific amino acid in the polypeptide. Translation takes place when the codons on mRNA bind to the *anticodons* on transfer RNA (tRNA) by complementary base pairing, where adenine binds with uracil and guanine binds with cytosine. Each tRNA molecule has a specific amino acid attached to it, which corresponds to its anticodon. The process takes place via the following steps:

Internal link

Ribosomes are discussed in **1.1 Cell structure and function**.

DP link

Deducing the DNA base sequence for the mRNA using a genetic code table will be studied in **2.7 DNA replication, transcription and translation** in the IB Biology Diploma Programme.

1. mRNA binds to the ribosome.
2. Translation begins at the "start" codon.
3. A tRNA molecule binds to the ribosome where its anticodon complements the codon on the mRNA.
4. Another tRNA molecule then bonds to the ribosome.
5. The two amino acids of the two tRNA molecules attach to each other via a peptide bond.
6. The first tRNA molecule separates from the ribosome, and the second tRNA molecule attaches to this position on the ribosome in its place.
7. On the mRNA, the ribosome moves to the next codon to allow another tRNA to bind.
8. The amino acid of the new tRNA binds to the chain of amino acids via a peptide bond.
9. The process continues until a "stop" codon is reached, where translation is stopped.
10. This process results in the formation of a *polypeptide* (amino acid chain) which is the basic structure of all proteins.

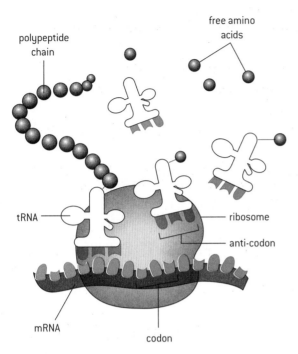

Figure 27. DNA translation

DP ready   Nature of science

Looking for patterns, trends and discrepancies
Most organisms use the 20 amino acids to make their proteins.
However, some organisms have used other amino acids to make
a polypeptide by special mechanisms. These were explained by
scientists as **variations** rather than a falsification of the theory that
20 amino acids are used to make proteins.

Chapter summary

In this chapter, you have learned about the biological molecules found in living organisms. You have
focused on the essential elements for life, particularly carbon, the structure of water and its function,
and the structure and function of the carbon compounds found in the human body including
carbohydrates, lipids, proteins and nucleic acids. Before moving on, please make sure that you have a
working knowledge of the following concepts and definitions:

☐ Some essential elements for life include hydrogen, carbon, oxygen and nitrogen.

☐ Carbon is the most essential element for life because it is required for the formation of most
biological molecules.

☐ Water consists of one oxygen atom and two hydrogen atoms held together by polar covalent
bonds.

☐ Water molecules are held together by hydrogen bonds, which contributes to the unique properties
of water.

☐ Carbohydrates are classified into three main categories: monosaccharides, disaccharides and
polysaccharides.

☐ Monosaccharides are the building blocks of all carbohydrates.

☐ Lipids include a wide range of hydrophobic compounds such as fats, oils, waxes, phospholipids
and steroids.

☐ Fats or triglycerides are made of one glycerol and three fatty acids linked together by covalent
bonds.

- ☐ Fatty acids include saturated fatty acids, cis unsaturated fatty acids and trans unsaturated fatty acids.
- ☐ Proteins perform essential functions in living organisms and they are composed of amino acids linked together by peptide bonds in either simple structures made of one chain or complex structures made of several chains.
- ☐ Enzymes are an important type of protein. They act as catalysts by speeding up chemical reactions, but do not get used up in the process.
- ☐ Enzymes are specific, meaning that each enzyme can only catalyse one reaction. This is because only a substrate molecule with a specific shape can bind to the active site of the enzyme.
- ☐ Enzymes are affected by factors such as temperature, pH and substrate concentration.
- ☐ Nucleotides are the building blocks of nucleic acids. Each is made of three main parts: a phosphate group, a pentose sugar and a nitrogenous base—in DNA the bases are adenine, guanine, thymine, or cytosine; in RNA uracil replaces thymine.
- ☐ Nucleotides in each DNA strand are joined together by covalent bonds while the two strands are held together by hydrogen bonds.
- ☐ DNA replication is the process by which DNA is copied inside the nucleus. Helicase and DNA polymerase are needed for the DNA replication process.
- ☐ Protein synthesis involves two processes: DNA transcription and translation.
- ☐ DNA transcription takes place in the nucleus, whereas DNA translation takes place in the cytoplasm.
- ☐ Each mRNA molecule carries a genetic code, where a triplet of bases forms a codon. Each codon codes for a particular amino acid. Codons will code for the same amino acids in all living organisms.

Additional questions

1. Explain the basic structure of a hydrocarbon. Give examples of hydrocarbons found in the human body.
2. Describe the bonding of carbon and hydrogen in methane.
3. Outline the structure of water.
4. Describe the functions of water in the human body.
5. Explain why lactose is converted to glucose and galactose in the food industry.
6. Explain the effect of pH on enzyme activity.
7. Explain why the structure of DNA is a double helix.
8. State the names of the enzymes involved in DNA replication and describe the role of each.
9. State the names of the molecules required for DNA translation and describe the role of each.

3 The human body

> " The human body is the only machine for which there are no spare parts. "
>
> **Herman Biggs**

Chapter context

The human body is a complex system that makes up the entire structure of a human being. It consists of different **organ systems** that work together to carry out specific functions necessary for life. The function of the **digestive system** is to break down large molecules into small molecules that can be absorbed into the blood stream. The **circulatory system** is responsible for transporting blood around the body. The **respiratory system** allows us to take in the vital oxygen and take out carbon dioxide. Through the **immune system**, our body fights against foreign particles such as bacteria and **viruses**. The **nervous system** controls our actions and responses. The **endocrine system** secretes **hormones** into the blood.

Learning objectives

In this chapter you will learn about the **structure** and **function** of the major organ systems including:

→ the digestive system

→ the circulatory system

→ the respiratory system

→ the immune system

→ the nervous system

→ the endocrine system.

🔑 Key terms introduced

→ Digestion, absorption and assimilation

→ Proteases, carbohydrases and lipases

→ Erythrocytes and leukocytes

→ Arteries, veins and capillaries

→ Intercostal muscles, antagonistic muscle action and the diaphragm

→ Aerobic and anaerobic respiration

→ Viruses

→ Lymphocytes, antigens and antibodies

→ The central nervous system (CNS) and peripheral nervous system (PNS)

→ Hormones

→ Homeostasis

→ Ovulation, follicle and oocyte

3.1 The digestive system

The *digestive system* is responsible for breaking down the large and insoluble food molecules into small and soluble molecules so that they are absorbed into the bloodstream. This involves the following processes:

- *Digestion*: The breakdown of the large and insoluble food molecules (macromolecules) into small and soluble molecules (monomers). Digestion takes place in different parts of the digestive system including the mouth, stomach and small intestine.

- *Absorption*: The transport of the small soluble products of digestion (monomers) into the bloodstream through the villi of the small intestine.

- *Assimilation*: The movement of the small soluble products of digestion by blood into the body tissues to be used by cells (figure 1).

🔑 Key term

The **digestive system** describes organs of the alimentary canal where the processes of digestion, absorption and assimilation occur.

🔑 Key term

Digestion is the breaking down of large, insoluble molecules into small, soluble molecules.

Absorption is the transport of soluble molecules (products of digestion) out of the intestines into the bloodstream.

Assimiliation is movement of the soluble products of digestion from the blood into the body tissues for use by cells.

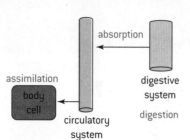

Figure 1. Digestion, absorption and assimilation

Internal link

The structure and function of the villi is discussed at the end of this section (**3.1 The digestive system**).

Key term

Proteases are enzymes that break down proteins into amino acids.

Carbohydrases are enzymes that break down carbohydrates into monosaccharides.

Lipases are enzymes that break down lipids into fatty acids and glycerol.

Internal link

Enzymes were introduced in **2.4 Proteins and enzymes**.

Key term

The **alimentary canal** is the long tubular organ where the processes of digestion take place. It is divided into discrete regions: mouth, pharynx, esophagus, stomach, small intestine, large intestine and anus.

Question

1   Distinguish between digestion, absorption and assimilation.

Enzymes and digestion

Digestion is a hydrolysis reaction, which means it involves the splitting of large molecules into smaller ones. Enzymes are essential for digestion to take place as they are integral to this process of breaking down large macromolecules. The resulting smaller molecules that can be absorbed into our blood are known as *nutrients*. These include amino acids, fatty acids, cholesterol and monosaccharides as well as vitamins and minerals. There are three main types of digestive enzymes (figure 2):

*   *Protease*—an enzyme that breaks down protein molecules into their building blocks, amino acids. There are many proteases produced by the digestive system. The main two are pepsin and trypsin.

*   *Carbohydrase*—an enzyme that breaks down carbohydrate molecules into their building blocks, monosaccharides (glucose, galactose and fructose). For example, amylase breaks down starch into maltose, which is then broken down by maltase into glucose.

*   *Lipase*—an enzyme that breaks down fat molecules into their building blocks, fatty acids and glycerol.

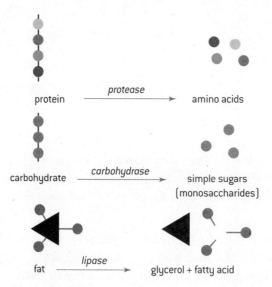

Figure 2. Digestive enzymes

The digestive system is a group of hollow organs that are connected together to form a long tube inside the body known as the *alimentary canal*. The alimentary canal includes the mouth, pharynx, esophagus, stomach and small and large intestines. In addition, there are three associated organs that help in the digestion process and these are the liver, pancreas and gallbladder (figure 3).

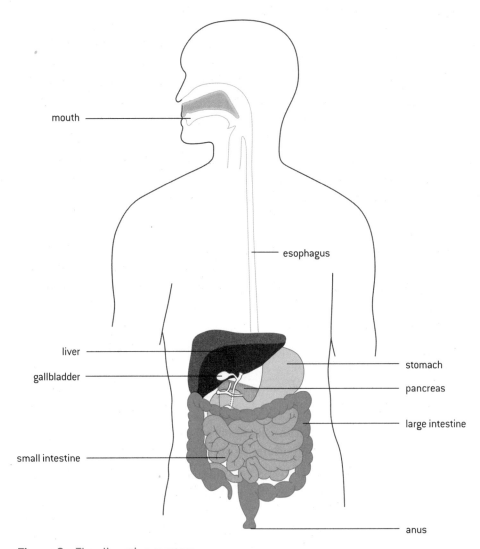

mouth

esophagus

liver

gallbladder

stomach

pancreas

large intestine

small intestine

anus

Figure 3. The digestive system

The mouth

The journey of food travelling through the alimentary canal starts in the mouth. There are two types of digestion that take place in the mouth: mechanical and chemical digestion.

Mechanical digestion involves the chewing of food by the teeth to make it small enough to be swallowed.

Chemical digestion involves the mixing of food in the mouth with the saliva, which contains enzymes, to break down food into smaller molecules.

Saliva is a watery substance that is secreted by the salivary glands, which are located near the teeth (figure 4). Saliva has several functions in the mouth:

- It moistens food so it can pass more easily from the mouth into the pharynx and esophagus.

- It contains the enzyme amylase, which breaks down starch into maltose. Maltose can be further broken down in the small intestine.

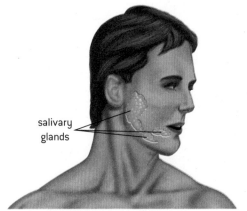

salivary glands

Figure 4. Salivary glands

Internal link

The respiratory system is discussed later in 3.3 Breathing and respiration.

Key term

Peristalsis is a wave of involuntary muscle contractions that pushes food along the alimentary canal.

Key term

A **sphincter** is a ring-shaped muscle that controls the opening and closing of a tube.

The pharynx and esophagus

The pharynx or throat is a passageway that extends from the mouth and nose into the esophagus and larynx (voice box). The pharynx has a role in moving the food into the esophagus. It also has a role in the respiratory system as it moves air from the nasal cavity into the larynx, and eventually into the lungs.

The esophagus is a muscular tube that extends from the pharynx to the stomach. The muscles of the esophagus contract in a wavy movement in order to move food down to the stomach. This wavy muscle contraction is called *peristalsis*.

The stomach

The stomach is a pear-shaped muscular sac that is located at the left side of the abdomen and connects the esophagus to the small intestine (figure 5). Both mechanical (by peristalsis) and chemical digestion (by enzymes) take place in the stomach. The stomach wall secrets digestive juices that contain:

*   Pepsin, the enzyme which begins the digestion of proteins into peptides and amino acids.
*   Hydrochloric acid (HCl), which kills bacteria and other harmful organisms, and provides the optimum pH for pepsin (pH 1.5–2).
*   Mucus, which forms a physical barrier that protects the surface of the stomach from being damaged by the hydrochloric acid.

The stomach is connected to the esophagus through the esophageal *sphincter* which prevents the acidic contents of the stomach from moving upward into the esophagus. The stomach is connected to the small intestine through the pyloric sphincter that opens to allow food to pass from the stomach to the small intestine.

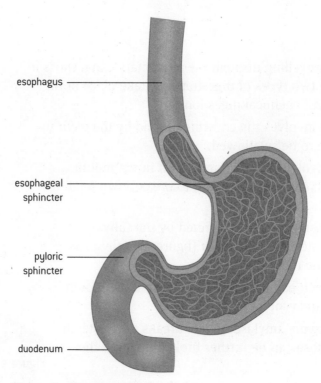

Figure 5. The stomach

The small intestine

The small intestine is a tightly folded long narrow tube about 6 m long where most digestion and absorption of food takes place (figure 6). It consists of three main parts: the duodenum (first section), the jejunum (middle section) and the ileum (final section).

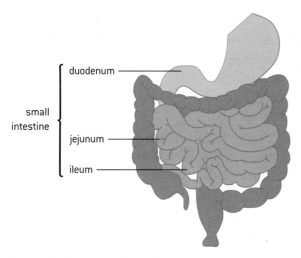

Figure 6. The small intestine

The small intestine is responsible for two main functions: completing digestion and absorption.

1. **Completing digestion**. Three important digestive juices are added to the food in the small intestine to complete digestion. These juices are:

   - Pancreatic juice, which is secreted by the pancreas and transferred via a duct to the small intestine. Pancreatic juice contains enzymes such as lipase (breaks down fats into glycerol and fatty acids), amylase (breaks down starch to maltose) and protease (breaks down proteins to amino acids). It also contains bicarbonate ions which are alkaline and neutralize stomach acids in the duodenum to maintain the pH between 7 and 8.

   - Bile, which is produced by the liver and stored in the gallbladder until release. It enters the small intestine through the bile duct. It helps in the digestion of lipids as it emulsifies fats. This means that it breaks down large drops of fats into small droplets, and therefore increases the surface area of the fats for the enzyme lipase to act upon.

   - Intestinal juice, which is secreted by glands in the wall of the small intestine. Intestinal juice contains carbohydrases, lipases and proteases to complete digestion. The digestive enzymes in the small intestine are immobilized on the *epithelial membrane*, which prevents the enzymes from being removed from the body.

2. **Absorption**. During absorption, the products of digestion (monomers), vitamins and minerals are absorbed into the blood stream via the *villi* present in the ileum. Villi are finger-like projections found on the inner surface of the ileum and they increase the surface area available for absorption. The structure of the villi is adapted to enable them to carry out absorption effectively (figure 7):

- The epithelium of each villus is only one cell layer thick (single-celled layer) and so allows for rapid absorption of food by facilitated diffusion and active transport.
- There are microvilli on the surface of the villi to further increase the surface area for absorption.
- Protein channels are present in the microvilli to allow facilitated diffusion. Pumps are also present for active transport. These structures increase the rate of absorption.
- Many mitochondria in the epithelium provide the ATP needed for active transport.
- Blood capillaries are very close to the epithelium so the diffusion distance is small.
- A lacteal is a lymphatic vessel at the centre of the villus which absorbs fats.

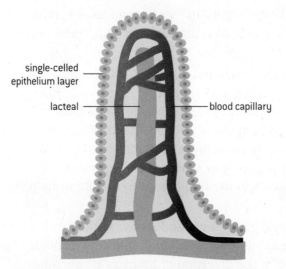

single-celled epithelium layer

lacteal

blood capillary

Figure 7. Structure of an intestinal villus

DP ready   Nature of science

The use of models can sometimes help us understand how a part of a living system may function. A simple example is the use of dialysis tubing, which is made from cellulose, to model the wall of the human small intestine. Pores in the dialysis tubing allow water and small molecules to pass through by osmosis and simple diffusion. Large molecules are too large to pass through. This is very similar to the lining of the small intestine where absorption takes place.

Practical skills: Modelling absorption in the small intestine using dialysis tubing

Dialysis tubing, which is also known as Visking tubing, is a partially permeable membrane that can be used to model absorption in the small intestine (figure 8). It could be used as follows:

- Wash the dialysis tubing with warm water to soften it and tie a knot in one end.
- Fill the dialysis tubing with starch solution and amylase solution. Then tie a knot in the other end and suspend it in a beaker of water.
- Amylase breaks down starch into maltose which is small enough to diffuse out of the dialysis tubing and into the beaker.
- Test the water for the presence of maltose by using Benedict's solution.

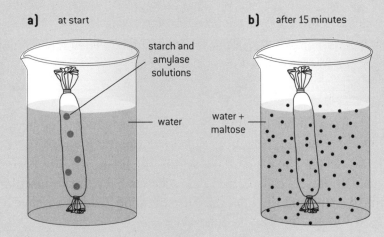

a) at start — starch and amylase solutions, water

b) after 15 minutes — water + maltose

Figure 8. Dialysis tubing (a) at the start of dialysis (b) after 15 minutes

The large intestine

The large intestine is the last part of the digestive system in humans. It is 1.5 metres long, which makes it shorter but thicker than the small intestine. The large intestine consists of three main areas: the cecum, colon and rectum. The main functions of the large intestine include absorption of water and storing the undigested food as feces in the rectum, which are then removed through the anus by defecation. The large intestine is also responsible for the absorption of vitamin K. Vitamin K is produced by the bacteria living inside the large intestine and is needed for blood clotting.

Question

2 Describe the role of enzymes in digestion with reference to two named examples.

3.2 Blood and circulation

Blood is circulated around the body in tubes called blood vessels. The heart is the organ that pumps blood throughout the body. The heart, blood and blood vessels make up the *circulatory system*. The function of the circulatory system is to transport oxygen and nutrients to all body cells and carry away waste products.

Key term

The **circulatory system** comprises the heart, blood vessels and blood. It transports oxygen and nutrients to all the body cells and removes waste products.

Internal link

Hormones will be discussed in **3.6 The endocrine system, hormones and homeostasis.**

 Key term

Erythrocytes are red blood cells. The name "red blood cell" derives from the appearance of the red blood cells after centrifuging as a red layer. The term erythrocyte comes from an ancient Greek origin where prefix "erythro" means "red" and "cyte" means "cell".

Leukocytes are white blood cells. The name "white blood cells" derives from the appearance of the white blood cells as a white layer after centrifuging. The term leukocyte comes from an ancient Greek origin where prefix "leuko" means "white" and "cyte" means "cell".

Key term

Arteries are the largest blood vessels; they carry blood away from the heart.

Veins are smaller than arteries, they carry blood to the heart.

Capillaries are the smallest blood vessels, they connect the arterioles and venioles.

Blood

Blood is a transport fluid that delivers necessary substances to cells including oxygen, and nutrients such as amino acids, and carries away waste products such as carbon dioxide and urea.

Blood is composed of the following:

- Plasma, which is a clear, yellowish liquid made up mainly of water. It also contains substances as proteins, *hormones*, salts and sugars.
- Red blood cells (*erythrocytes*), which make up around 40% of blood volume. Red blood cells are formed in the bone marrow. Their life span is about 120 days and therefore must be constantly formed. Their main function is to transport oxygen to cells and carry away carbon dioxide.
- White blood cells (*leukocytes*), which have major role in the body's immune response. They are essential to protect the body against foreign particles such as bacteria and *viruses*.
- Platelets, which are the smallest cells in blood. They have a major role in blood clotting and control of bleeding.

Blood vessels

Blood vessels are tubes in which blood circulates around the body. There are three main types of blood vessels (figures 9 and 10):

- *Arteries*, which carry blood away from the heart. Arteries branch into small vessels called arterioles. Arteries have thick muscular walls to withstand the high pressure of blood. They also have a narrow lumen (internal passageway).
- *Veins*, which carry blood to the heart. Veins branch into small vessels called venules. Veins have thinner walls and a larger lumen than arteries. The blood they contain is at low pressure. Veins also have valves to prevent the backflow of blood.
- *Capillaries*, which are the very thin vessels that connect the arterioles and the venules. They are one cell thick to allow for the exchange of materials between cells in the tissue and the blood in the capillary. For example, oxygen diffuses from the capillary into the tissue and carbon dioxide diffuses from the tissue into the capillary.

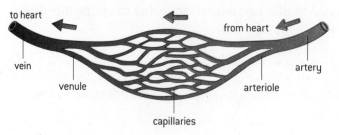

Figure 9. Blood vessels

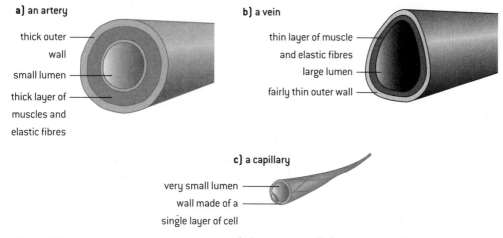

Figure 10. Cross-sectional structure of **(a)** an artery, **(b)** a vein and **(c)** a capillary

Question   Ⓠ
3   Distinguish between the structures of the three blood vessels.

Heart

The heart is a muscular organ that is about the size of a closed fist (figure 11). The heart is located in the middle of the chest between the two lungs. Its function is to pump blood throughout the body.

The heart has the following structure:

- It contains four chambers, including the right atrium, left atrium, right ventricle and left ventricle. The atria are smaller and thinner than the ventricles.

- It contains four valves that prevent the backflow of blood. Two atrioventricular valves between the atrium and the ventricle on each side. Two semilunar valves at the entrance of the pulmonary artery and aorta.

- The left side of the heart has thicker walls than the right side of the heart to withstand the high pressure of blood in the right side.

- It is connected to four main blood vessels:
  - → Pulmonary vein—carries oxygenated blood from the lungs to the heart
  - → Aorta—carries oxygenated blood from the heart to body cells
  - → Vena cava—carries deoxygenated blood from body cells to the heart
  - → Pulmonary artery—carries deoxygenated blood from the heart to the lungs.

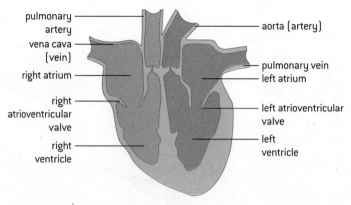

Figure 11. Structure of the heart

Blood circulation (double circulation)

The left side of the heart pumps oxygenated blood around the body (systemic circulation), while the right side of the heart pumps deoxygenated blood to the lungs (pulmonary circulation). The two sides of the heart are separate so that the oxygenated blood on the left side does not mix with deoxygenated blood on the right side (figure 12).

Blood circulation is described in the following steps:

1. Both atria collect blood from veins. The left atrium collects oxygenated blood from the pulmonary vein while the right atrium collects deoxygenated blood from the vena cava.

2. Both atria contact and force blood into the ventricles through the atrioventricular valves which open due to the pressure exerted by blood.

3. Then the ventricles contract and force blood out into arteries through the semilunar valves. The atrioventricular valves close to prevent backflow of blood into the atria.

4. Arteries carry blood away from the heart. The aorta carries oxygenated blood from the heart to body cells while the pulmonary artery carries deoxygenated blood from the heart to the lungs.

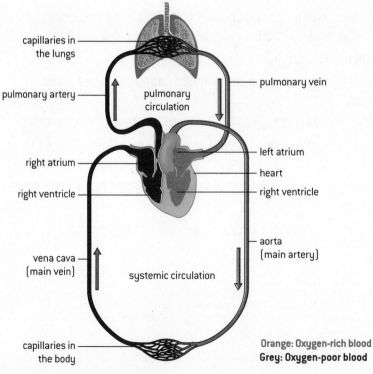

Figure 12. Double circulation

Theories can never be fully proved, only disproved. William Harvey (figure 13), an English physician, overturned Galen's theories on movement of blood in the body. According to the ancient Greek philosopher Galen, the liver makes blood for the veins and the heart makes blood for the arteries. Through observation and experimentations, Harvey explained the double circulation of the heart where blood circulates from the heart to the lungs, and from the heart to the rest of the body.

Figure 13. William Harvey

Systole and diastole

Systole is when the heart muscle is contracting to push blood out of any of the chambers. This takes place when the atria contract to force blood into the ventricles, and when the ventricles contract to force blood into the arteries to be carried away from the heart.

Diastole is when the heart muscle is relaxing. This takes place when blood is moving into the atria from veins.

Coronary arteries

The coronary arteries are branches of the aorta which supply the heart muscle with oxygen and nutrients (figure 14).

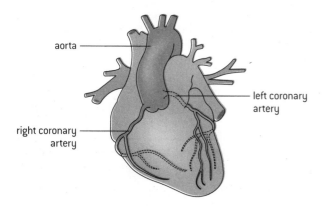

Figure 14. Coronary arteries

The coronary arteries may become blocked which reduces the supply of oxygen to the heart. This may cause coronary heart disease (CHD).

DP link

Systole and diastole are to be studied in **6.2 The blood system** in the IB Biology Diploma Programme.

Key term

Atherosclerosis is the hardening and narrowing of the arteries due to the deposition of cholesterol.

When do coronary arteries get blocked?

Cholesterol can stick to the walls of the arteries, causing them to become hard and rough. This is called *atherosclerosis*. Fatty deposits, cholesterol, plaque and platelets block the artery and form a clot (thrombus) that prevents blood flow (figure 15). If a coronary artery is blocked, a section of the heart does not get enough nutrients and oxygen, which can cause a heart attack.

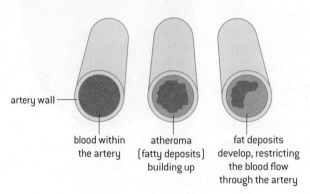

artery wall

blood within the artery

atheroma (fatty deposits) building up

fat deposits develop, restricting the blood flow through the artery

Figure 15. Blockage of coronary arteries

There are some risk factors that are believed to increase the chance of having CHD. These risk factors may include smoking, obesity, diabetes, a diet with a high intake of saturated fat, lack of exercise, high blood pressure, genetics and gender (men are more likely to have CHD than women).

Practical skills: Dissecting a sheep's heart

In your DP Biology class, you might perform a dissection of a sheep's heart, to identify the chambers and valves of the heart and the blood vessels connected to it. Do not perform this experiment without adequate safety precautions.

A typical procedure for this dissection is as follows:

* Identify the right and left sides of the heart. The side that has the pointed end is the left side. The left side will feel much stronger than the right side. Notice the difference in the thickness of the wall for both sides.
* Examine the openings at the top including the aorta and vena cava.
* Cut open the heart.
* Examine the inside of the heart and look for the valves between the atria and ventricles and the valves at the base of the pulmonary artery and aorta. Also look for the coronary arteries.
* Examine the four chambers of the heart.

Control of the heartbeat

The normal heart beats around 70 beats per minute. So, what controls our heartbeats?

The heartbeat is controlled by the *pacemaker,* which is a group of specialized muscle cells located in the right atrium. The pacemaker sends electrical signals that cause the heart to contract.

The pacemaker receives impulses from the brain to increase or decrease the heart rate. During exercise, the heart rate increases because the body needs more oxygen for cell respiration.

Key term

The **pacemaker** is a group of specialized muscle cells located in the right atrium and is responsible for the control of the heartbeat.

Question

4 Explain how the heartbeat is controlled.

5 Outline the four main blood vessels connected to the heart and describe their functions.

3.3 Breathing and respiration

Breathing is the process by which the body moves oxygen into the lungs and carbon dioxide out of the lungs. This process is also known as ventilation. The difference in concentration of oxygen and carbon dioxide between the lungs and the blood in capillaries allows for gas exchange to take place where oxygen diffuses from the lungs into the blood in capillaries and carbon dioxide diffuses from the blood in capillaries into the lungs.

Respiration is the production of energy (in the form of ATP) in living cells and thus named cellular respiration. There are two types of cellular respiration: aerobic and anaerobic. In aerobic respiration, oxygen is required for the production of ATP and carbon dioxide is released as a waste product. Anaerobic respiration takes place in the absence of oxygen.

The respiratory system

The main function of the *respiratory system* is to bring in oxygen which is needed for aerobic respiration and get rid of carbon dioxide which is a waste product. The respiratory system consists of several organs that are involved in breathing. These organs include the nose, pharynx, larynx, trachea, bronchi and lungs (figure 16).

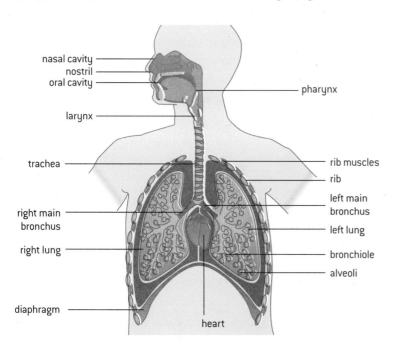

Figure 16. Structure of the respiratory system

When you breathe in (inhale), air containing oxygen enters through your nose or mouth and passes through the pharynx, the larynx (the voice box) and then the trachea to get into your lungs through one of the two bronchi. Each bronchus is branched into many bronchioles and each bronchiole ends in millions of tiny sacs called alveoli.

DP link

Control of the heartbeat and factors influencing the rate of heartbeat are to be studied in **6.2 The blood system** in the IB Biology Diploma Programme.

Key term

Breathing (ventilation) is the process by which the body gets oxygen into the lungs and carbon dioxide out of the lungs.

Respiration (also know as cellular respiration) is the production of energy (ATP) in living cells.

Internal link

Aerobic and anaerobic respitation will be discussed later on in this section (3.3 Breathing and respiration).

Key term

The **respiratory system** provides the oxygen required for aerobic respiration and removes carbon dioxide. It consists of a number of organs: the nose, pharynx, larynx, trachea, bronchi and lungs.

The alveoli are the site of gas exchange, where oxygen diffuses out of the alveoli into the blood in capillaries and carbon dioxide diffuses out of the blood in capillaries into the alveoli (figure 17).

The structure of the alveoli enables them to carry out gas exchange. There are a huge number of alveoli in each lung and this increases the surface area for gas exchange. The alveoli are adapted as follows:

- The wall of the alveolus is made up of a single layer of thin cells. This allows for rapid gas exchange.

- Each alveolus is surrounded by a network of blood capillaries. This allows gases to diffuse quickly into and out of blood.

- The alveoli are moist. This allows gases to dissolve before gas exchange takes place.

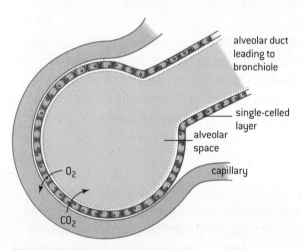

Figure 17. The mechanism of the alveolus

Breathing mechanism

As we inhale or exhale, our ribcage and chest move to force air in and out (figure 18). The movement of the chest is due to the contraction and relaxation of several muscles that help in the breathing process:

- The *intercostal muscles* are a group of muscles located between the ribs. They include two types: internal and external intercostal muscles. The two types of muscles work oppositely so that when one muscle contracts the other relaxes. This type of movement is called *antagonistic muscle action*.

- The *diaphragm* is a C-shaped muscle sheet located at the bottom of the thorax, below the lungs.

As we breathe in (inhale):

- The internal intercostal muscles relax while the external intercostal muscles contract. This causes the ribcage to move upwards and outwards.

- The diaphragm contracts, flattens and moves downward. This causes the volume of the thorax to increase.

- The pressure of the thorax decreases, and air is forced into the lungs.

As we breathe out (exhale):

- The internal intercostal muscles contract while the external intercostal muscles relax. This causes the rib cage to move downwards and inwards.

 Key term

Intercostal muscles are a group of paired muscles located between the ribs.

Antagonistic muscle action describes the movement of muscles that work in pairs; when one muscle contracts, the other relaxes.

The **diaphragm** is C-shaped sheet of muscle located at the bottom of the thorax, below the lungs.

- The diaphragm relaxes, curves and moves upward. This causes the volume of the thorax to decrease.
- The pressure of the thorax increases, and air is forced out of the lungs.

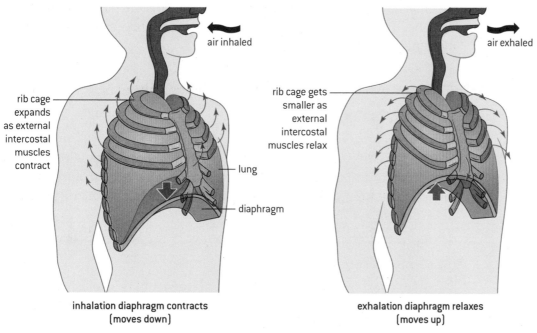

inhalation diaphragm contracts (moves down)

exhalation diaphragm relaxes (moves up)

Figure 18. The breathing mechanism

Practical skills: Monitoring of breathing in humans

Breathing (ventilation) can either be monitored by measuring the following:

1. **Ventilation rate** = number of times air is inhaled or exhaled per minute. It can be measured by:
   - Simple observation—count the number of times air is exhaled or inhaled in a minute.
   - Using a data logger with a spirometer, or chest belt and pressure meter.

2. **Tidal volume** = volume of air inhaled or exhaled in a normal breath. It can be measured by:
   - One normal breath is exhaled through a tube into a vessel and volume is measured (figure 19).
   - Using a data logger and a spirometer.

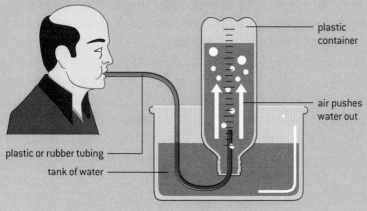

plastic container

air pushes water out

plastic or rubber tubing

tank of water

Figure 19. Measuring tidal volume

Cellular respiration

Cellular respiration is the process by which energy is released from organic compounds such as glucose or other molecules (such as fatty acids or amino acids) in living cells. Energy is released in the form of ATP.

Adenosine triphosphate (ATP) is the energy-carrying molecule used in cells. It releases energy when it loses one phosphate group (P_i) and becomes adenosine diphosphate (ADP).

$$ATP \rightarrow ADP + P_i$$

There are two types of respiration: aerobic and anaerobic.

Aerobic respiration

Aerobic respiration is the reaction of glucose and oxygen to produce carbon dioxide, water and a large amount of energy. It starts in the cytoplasm of the cell but continues in the mitochondria. It results in the production of a large yield of ATP, as the energy produced from this reaction is used to convert ADP to ATP.

glucose + oxygen → carbon dioxide + water (+ energy to produce ATP)

Anaerobic respiration

Anaerobic respiration does not require oxygen and is also called fermentation. It occurs in the cytoplasm and gives a small yield of ATP from glucose (or other organic molecules).

In humans, anaerobic respiration is known as lactate fermentation, and usually occurs during exercise when the respiratory system is unable to bring into the body enough oxygen to the muscles to respire aerobically. It produces lactic acid (lactate), which causes muscle fatigue and pain. The reaction is usually reversible, which means that when oxygen is available, lactic acid will be converted back to continue with aerobic respiration.

glucose → lactic acid (+ energy to produce a small yield of ATP)

In yeast, anaerobic respiration is usually called alcoholic fermentation, as it releases ethanol (a type of alcohol) and carbon dioxide. This reaction is important in making bread, as the carbon dioxide forms bubbles in bread, causing the dough to expand and rise. The ethanol evaporates during baking.

glucose → ethanol + carbon dioxide (+ small yield of ATP)

Practical skills: Investigate factors affecting the rate of yeast fermentation

You can do an experiment to investigate any of the factors that may affect the rate of yeast fermentation. One such factor is the amount of glucose available for fermentation.

An experiment you may do in class in the Diploma Programme would involve preparing different concentrations of glucose solution and then adding yeast. You would measure the volume of carbon dioxide produced in a given period of time to determine the average rate of fermentation. The average rate of fermentation can be calculated by dividing the volume of carbon dioxide produced by time.

$$\text{Average rate of fermentation} = \frac{\text{volume of CO}_2 \text{ produced}}{\text{time}}$$

You also need to keep many variables constant, such as the volume and temperature of the glucose solution, and the mass of yeast used.

Worked example: Effect of temperature on fermentation rate

1. A student conducted an experiment to investigate the effect of changing temperature on the rate of yeast fermentation by measuring the volume of carbon dioxide produced in five minutes. The following results were obtained (table 1):

Table 1. Effect of changing temperature on the volume of carbon dioxide produced over a five minute period

Temperature (°C)	Volume of carbon dioxide produced (cm³)
20	2.0
25	3.5
30	5.0
35	1.5
40	0.5

a) What are the independent and dependent variables?

b) What are the variables that may have been controlled?

c) How can you comment on the reliability of the data obtained?

Solution

a) The independent variable is temperature and the dependent variable is the volume of carbon dioxide produced.

b) Variables that may have been controlled include mass of yeast, volume of sugar solution, volume of water and yeast brand.

c) In scientific investigations, repeating experiments increases the reliability of data obtained. As seen from table 1, there were no repeats in this experiment, which reduces the reliability of data obtained. The experiment must be repeated as many times as possible to reach a more reliable conclusion.

2. Using the data in table 1, calculate the average rates of fermentation for each temperature.

Solution

Using the equation:

$$\text{Rate of fermentation} = \frac{\text{volume of } CO_2 \text{ produced}}{\text{time}}$$

At 20°C, the rate of fermentation $= \dfrac{2.0}{5.0} = 0.4 \text{ cm}^3 \text{ min}^{-1}$

Table 2 shows the average rate of yeast fermentation for all the temperatures investigated by the student.

Table 2. Rate of yeast fermentation at different temperatures

Temperature (°C)	Volume of carbon dioxide produced (cm³)	Rate of yeast fermentation (cm³ min⁻¹)
20	2.0	0.40
25	3.5	0.70
30	5.0	1.0
35	1.5	0.30
40	0.5	0.10

Practical skills: Using a respirometer

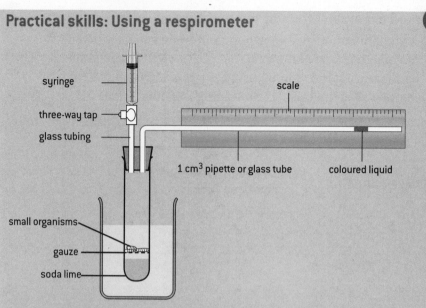

Figure 20. The respirometer

It is possible to measure a living organism's rate of respiration. This is done by using a respirometer, which measures the rate of exchange of oxygen and/or carbon dioxide (figure 20). A simple respirometer measures the volume of oxygen consumption of an organism over a specific time. An alkali, such as soda lime, is added to the apparatus to absorb CO_2, thus any reductions in volume will only be due to oxygen consumption by the organism.

A drop of coloured liquid is inserted into the glass tube. To determine the rate of respiration, the distance travelled by the coloured drop is measured at one minute intervals for a given amount of time.

DP ready   Nature of science

Ethical implications of using invertebrates in respirometer experiments

The use of invertebrates in experiments caused a huge debate about the ethical implications of using animals in experiments. Before carrying out an experiment involving living organisms, we should consider whether the experiment is ethical—particularly if the experiment is causing any harm, pain or distress to the animal.

Question

6 Compare and contrast between aerobic and anaerobic respiration.

7 Distinguish between breathing and cellular respiration.

3.4 Body defence

The *immune system* describes the various ways in which our body fights disease. There are three lines of defence that protect the human body against *pathogens* that may enter the body. The first line of defence includes the natural barriers such as the skin and mucous membranes. The second line of defence includes the non-specific immune response by phagocytes. The third line of defence includes the specific immune response by *lymphocytes*. Lymphocytes are a specific type of white blood cell (or leukocyte).

DP link

The guidelines for the use of animals in your practical work on the DP Biology course will be discussed in the animal experimentation policy.

Key term

The **immune system** describes the various ways in which our body fights disease.

What are pathogens?

A pathogen is anything that causes a disease. Pathogens may attack cells of humans, animals or even plants to cause a disease. Pathogens include:

- Bacteria (example of disease: tuberculosis)
- Viruses (example of disease: hepatitis B)
- Fungi (example of disease: athlete's foot)
- Protists (example of disease: malaria).

What are viruses?

Viruses are small infectious agents that are considered non-living organisms. However, viruses replicate and reproduce inside a living host cell, transforming the cell into a virus-making factory.

All viruses are made up of:

- A genetic material, either DNA or RNA
- A protein coat.

However, viruses vary in shape, structure and the type of genetic material they contain (figure 21).

HIV   bacteriophage   influenza

Figure 21. Examples of viruses

How are pathogens transmitted?

Transmission of pathogens into humans takes place through different routes. Table 3 shows some ways of transmitting pathogens and an example of each.

Table 3. Transmission of pathogens

Route of transmission	An example of a disease
Blood	Hepatitis B
Droplets in the air	Flu
Food	Food poisoning by salmonella
Sexual intercourse	HIV
Water	Cholera
Via animals such as insects	Malaria

Treatment of infectious diseases

Some drugs may kill the infectious pathogen. Antibiotics are drugs used to kill bacteria but not viruses. This is because antibiotics interrupt the processes that take place in prokaryotic cells, such as bacteria. Antibiotics work by different methods such as:

- Breaking down the cell wall
- Stopping protein synthesis
- Stopping DNA replication.

Key term

A **pathogen** is any organism or virus that causes a disease, such as bacteria, viruses and fungi.

Key term

Viruses are non-living things that reproduce inside a living host cell.

Penicillin was the first antibiotic, discovered by Alexander Fleming in 1928 (figure 22). The discovery of penicillin was an accident, resulting from the unintended contamination of a dish containing *Staphylococcus aureus*. A *Penicillium* mould began to grow on the plate and a zone of inhibited bacterial growth was observed around the mould. Fleming concluded that the mould was releasing a substance (penicillin) that was killing the nearby bacteria.

Question

8 Explain why antibiotics are effective against bacteria but not viruses.

Figure 22. Alexander Fleming

The first line of defence

The first line of defence includes the natural barriers that protect the body from the outside pathogens. These natural barriers include skin, stomach acid and mucous membranes (figure 23).

Skin forms a physical barrier that prevents pathogens from entering the body unless it is damaged. Skin could be damaged due to burns or injuries. The skin also contains glands that secrete some acids which prevent the growth of pathogens on the surface.

Mucous membranes are the moist linings present at body openings such as the mouth, nose, trachea and vagina. The mucus present in such areas contains enzymes that breaks down pathogens. The respiratory system has sticky mucus which can trap dust and pathogens. Some mucous membrane tissues have tiny hair-like structures called cilia that can trap pathogens and remove them out of the body.

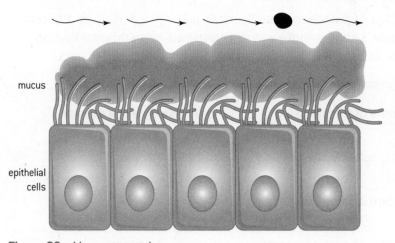

Figure 23. Mucous membranes

The hydrochloric acid (HCl) secreted by the stomach increases the acidity of the stomach and thus kills bacteria and other pathogens that might be in the food we eat.

If pathogens manage to pass through our natural barriers, then the immune system is stimulated to develop an immune response.

The second line of defence

The second line of defence involves the phagocytic white blood cells, which are also called macrophages. This is a non-specific immune response which means that the phagocytes will only recognize the pathogen as a non-self—something foreign to the body. Once they do, the following takes place:

 Key term

Phagocytosis is the process that occurs when a phagocyte engulfs (takes into the cell) bacteria or other material.

- Phagocytes engulf the pathogen by *phagocytosis* (figure 24).
- Organelles within the phagocytes called lysosomes then digest the pathogens. Lysosomes contain enzymes that help to kill these pathogens.

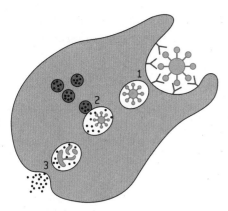

Figure 24. The process of phagocytosis: (1) The pathogen is engulfed by macrophage; (2) Lysosomes digest the pathogen; (3) The pathogen waste products are removed from the cell

The third line of defence

The third and the final line of defence involves the production of Y-shaped *antibodies* by lymphocytes in response to the presence of particular pathogens. Pathogens have proteins on their surfaces that stimulate the production of specific antibodies (figure 25). These proteins are called *antigens*.

 Key term

Lymphocytes are a type of white blood cells.

Antigens are proteins found on the surfaces of pathogens which stimulate the production of specific antibodies.

Antibodies are proteins that defend the body against pathogens by binding to the antigens found on the surface of these pathogens.

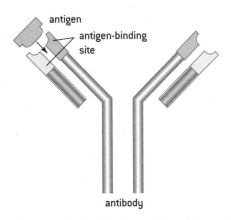

Figure 25. Antibody–antigen interaction

How are antibodies produced?

Some lymphocytes make a specific antibody for the antigen they recognize:

- Antibodies are found on the surface of the lymphocytes with the antigen-binding site projecting outwards.
- Pathogens have antigens on their surface.
- Each antibody recognizes a specific antigen.
- The antigens bind to the antigen-binding site of the antibodies of a specific lymphocyte (figure 26).
- The lymphocyte becomes active and starts to make more antibodies.
- Some cells remain in the blood stream waiting for a secondary infection to produce more antibodies and cause a faster response.

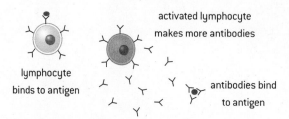

Figure 26. Antibody production

Question (Q)

9 Describe the production of antibodies to fight a pathogen.

Blood clotting

When you are injured or cut, blood clotting is a necessary process that is needed to stop you from losing too much blood. When skin is cut, the damaged skin and platelets release a specific type of protein at the wound site called clotting factors. Clotting factors convert an enzyme, called prothrombin, from its inactive form to its active form, which is called thrombin. Thrombin catalyses the conversion of a soluble protein (known as fibrinogen) into an insoluble fibrous protein (known as fibrin) which attaches to red blood cells to form a clot (figure 27).

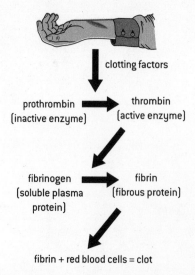

Figure 27. The process of blood clotting

3.5 The nervous system

The *nervous system* is a complex network of nerves and specialized cells that control your actions and coordinate your muscles.

The nervous system consists of two main parts (figure 28):

- The *central nervous system* (*CNS*) which consists of the brain and spinal cord.
- The *peripheral nervous system* (*PNS*) which consists of the nerve cells that connect all parts of the body with the CNS.

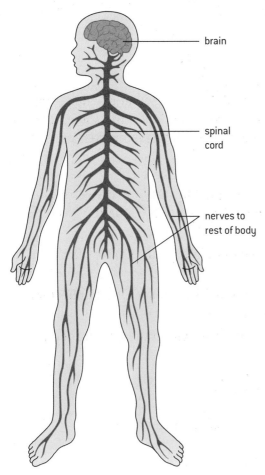

Figure 28. The nervous system

The brain

The brain is located at the top of the spinal cord and is protected inside the skull. The brain consists of three main parts (figure 29):

- Cerebral hemisphere—the part of the brain that controls complex behaviours such as thought, memory and learning.
- Medulla (also called medulla oblongata)—the part that attaches the brain to the spinal cord. It controls our autonomic functions such as breathing, digestion, heart rate and swallowing.
- Cerebellum—controls our voluntary movements such as balance, coordination and speech.

Internal link

The hypothalamus and pituitary glands will be studied later in **3.6 The endocrine system, hormones and homeostasis.**

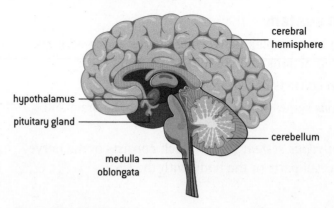

Figure 29. The brain

Spinal cord

The spinal cord is a long thin tube-like structure that extends from the medulla oblongata in the brain to the bottom of the spine. It is protected by the bones of the spine which are called vertebrae.

Nerve cells

The nerve cells, or **neurons**, are adapted to carry rapid electrical impulses from one area to another. There are three main types of neurons (figure 30):

- Sensory neurons—carry nerve impulses from receptors (sensory cells) to the CNS.
- Motor neurons—carry nerve impulses from the CNS to effectors (muscle and gland cells) (figure 31).
- Relay neurons—carry nerve impulses within the CNS from one neuron to another.

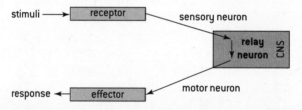

Figure 30. The main types of neurons

A receptor is a group of specialized cells that can detect a stimulus and produce an electrical impulse in response. This includes the receptors found in our sensory organs.

An effector is any part of the body that produce a response to a stimulus. This includes muscles and glands.

A neuron is like other cells as it consists of a cell membrane, cytoplasm and a nucleus, but its shape is very much different. It is stretched out to form a long part called the axon. The length of an axon can reach over a metre long. Some of the neurons are covered with a *myelin sheath*, which speeds up conduction of electrical impulses. Neurons also have tiny branches called dendrites.

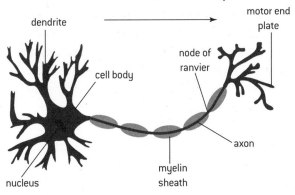

Figure 31. A motor neuron

How do electrical impulses travel?

Electrical impulses are collected by the dendrites of the neuron and then transferred through the axon to reach the end of the neuron where it is then passed to another neuron. Electrical impulses travel very fast along the axon of the neuron.

When the neuron is not sending an electrical impulse, it is "at rest". A neuron is at rest when the inside of the axon is negatively charged, relative to the outside of the axon. This is because of the active transport of sodium ions (Na^+) and potassium ions (K^+) through the sodium–potassium pump, in which three sodium ions move out of the axon and two potassium ions move into the axon. The inside of the axon develops a net negative charge as a result. The neuron develops an electrical potential across the plasma membrane, and does not send any electrical impulses. We say that the neuron is at *resting potential* (figure 32).

When the neuron is sending an electrical impulse, it is in "action". A strong stimulus may cause the sodium channels to open and sodium ions (Na^+) diffuse into the axon. This causes the inside of the axon to develop a net positive charge compared to the outside. The high positive charge inside the axon will cause the potassium channels to open and the potassium ions (K^+) diffuse out of the axon. This results in switching the potential back to its original configuration, where a net negative charge is developed inside the axon. We say that the neuron is at *action potential* (figure 32).

> ### Key term
>
> **Resting potential** is the electrical potential across the plasma membrane of a neuron that is not sending an impulse.
>
> **Action potential** is the electrical potential across the plasma membrane of a neuron where an impulse is passed through.

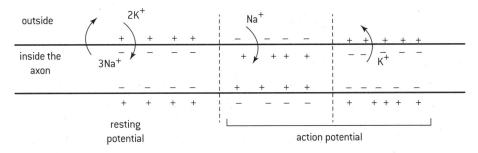

Figure 32. Resting and action potential in a neuron

Synapses

A synapse is the tiny gap that exists between two neurons (figure 33). When the electrical impulse reaches the end of a neuron, it causes the release of *neurotransmitters* into the synapse. Neurotransmitters are chemicals that send signals between neurons. The neurotransmitters pass across the synapse from one nerve to the other, and cause the following neuron to transmit an electrical impulse.

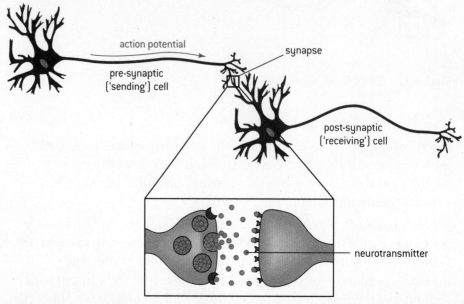

Figure 33. A synapse

DP link

The transfer of nerve impulse within the neuron and between synapses are to be studied in details in **6.5 Neurons and synapses** in the IB Biology Diploma Programme.

Key term

Hormones are chemical substances secreted by endocrine glands into bloodstream.

Question

10 Explain how the nerve impulse is passed within the neuron.

3.6 The endocrine system, hormones and homeostasis

The endocrine system consists of glands that secrete hormones which are transported in the blood (figure 34). Hormones travel in the blood to affect a target tissue, resulting in a response which modifies the internal environment.

Hormones affect different target tissues to cause a specific response. They help to regulate the functions of many cells and organs.

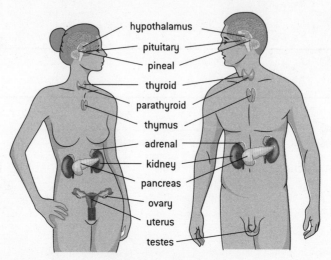

Figure 34. The endocrine system

Homeostasis

Homeostasis maintains the conditions of the body's internal environment, within the necessary range of values, for the organs of the body to function effectively. The internal environment consists of blood and tissue fluid. While the external environment often varies greatly, the body's internal environment varies very little, relying on *negative feedback* mechanisms to keep it within the normal range. The negative feedback mechanisms are controlled by the nervous and endocrine systems.

For example, changes in the levels of blood glucose concentration affect the rate at which blood glucose is produced. When the levels of blood glucose rise above the set point they trigger a decrease in the production of blood glucose causing the blood glucose concentration levels to fall until they reach the correct level. If the blood glucose concentration levels are reduced below the set point, the negative feedback system causes the production of blood glucose to increase until the concentration level returns to the set point. Because it is normal for the concentration levels of the various factors of the internal environment to fluctuate slightly, small fluctuations around the set point do not cause a feedback response. Negative feedback is only triggered when concentration levels are significantly above or below the set point.

Key term

Homeostasis is the ability of the body to maintain the conditions of its internal environment.

Negative feedback maintains the body's internal environment by stabilizing the effect of any change above or below the set point or range for factors such as blood glucose concentration or hormone regulation.

DP link

Hormones and their effects are to be studied in greater detail in **6.6 Hormones, homeostasis and reproduction** in the IB Biology Diploma Programme.

Question

11 What are the two systems that are involved in the control of homeostasis?

The hypothalamus and the pituitary gland

The hypothalamus is a small region of the brain located at the base of the brain near to the pituitary gland (figure 35). It has a very crucial role in controlling the endocrine system by linking it to the nervous system. The hypothalamus sends signals to the pituitary gland to release or inhibit the secretion of several hormones that control different functions in the body.

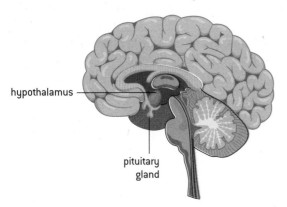

Figure 35. Hypothalamus and pituitary glands

Two of the most important hormones secreted by the hypothalamus are:

- Thyrotropin-releasing hormone (TRH), which stimulates the pituitary gland to produce the thyroid-stimulating hormone (TSH).
- Gonadotropin-releasing hormone (GnRH), which stimulates the pituitary gland to produce important reproductive hormones, such as follicle-stimulating hormone (FSH) and luteinizing hormone (LH).

Hormones secreted by the hypothalamus regulate body temperature, appetite, thirst, sleep and sex drive.

The most important hormones secreted by the pituitary gland are:

- Prolactin: It stimulates milk production in females.
- Thyroid-stimulating hormone (TSH): it regulates the body's thyroid gland and the secretion of thyroxine.
- Luteinizing hormone (LH): In women, it regulates estrogen. In men, it regulates testosterone.
- Follicle-stimulating hormone (FSH): Found in both men and women. It stimulates the releasing of eggs in women and helps ensure the normal function of sperm production in men.
- Oxytocin: causes contractions in pregnant women during child birth and also promotes milk flow in nursing mothers.

Pancreas

The pancreas acts as an endocrine gland. Insulin and glucagon are secreted by the β and α cells of the pancreas respectively, to control blood glucose concentration (table 4). Cells in the pancreas, called islets of Langerhans, monitor the concentration of blood glucose and send hormone messages to target organs when the level is low or high. This is done by negative feedback.

Table 4. Response to change in blood glucose levels

Response to high blood glucose levels	Response to low blood glucose levels
• Insulin is produced in the β cells in the pancreatic islets (islets of Langerhans).	• Glucagon is produced by the α cells in the pancreatic islets.
• Insulin targets liver cells and muscle cells.	• Glucagon targets liver cells and muscle cells.
• Insulin triggers glucose to be taken from the blood and converted into glycogen in the liver and muscle cells.	• In the liver cells and muscle cells glucagon triggers the conversion of glycogen back into glucose.
• Glycogen is then stored in the cellular cytoplasm.	• The liver cells and muscle cells release the glucose into the blood.
• In addition, glucose is taken up by other cells where it is used for cell respiration instead of fat.	• This raises the blood glucose concentration levels.
• All of these processes lower the blood glucose concentration levels.	

Thyroid gland

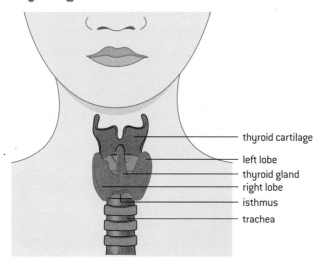

Figure 36. The thyroid gland

The thyroid gland secretes thyroxine under the influence of thyroid-stimulating hormone (figure 36). Thyroxine contains iodine and exist in two forms: T_3 and T_4. It increases metabolic rate and body temperature. Thyroxine levels are maintained within narrow limits by negative feedback.

Hyperthyroidism occurs when there is too much thyroxine in the body.

Hypothyroidism occurs when there is too little thyroxine in the body. Symptoms of hypothyroidism include weight gain, loss of energy, feeling cold all the time, forgetfulness and depression. Goitre may take place (enlargement of the thyroid). This is because of prolonged deficiency of iodine in the diet which prevents the formation of thyroxine.

Reproductive hormones

Three of the most important reproductive hormones are testosterone, progesterone and estrogen. In males, testosterone is the significant reproductive hormone. Progesterone and estrogen are the significant reproductive hormones in females.

Testosterone in males

A gene on the Y chromosome causes the development of the testes which secrete testosterone.

Testosterone has the following functions:

1. Pre-natal development of male genitalia
2. Sperm production
3. Development of male secondary sexual characteristics during puberty.

Progesterone, estrogen and the female reproductive system

There are a number of hormones involved in female reproduction. They are progesterone, estrogen, follicle-stimulating hormone (FSH), luteinizing hormone (LH) and gonadotrophin-releasing hormone (GnRH).

Progesterone and estrogen are produced by the ovaries under the influence of FSH and LH, which are both produced in the pituitary gland.

The hypothalamus produces gonadotrophin-releasing hormone (GnRH), which targets the pituitary gland to produce FSH and LH.

 Internal link

Genes and sex chromosomes are to be studied later in **5.1 The material of inheritance** and **5.3 Inheritance**.

Both FSH and LH target the ovaries to control the menstrual cycle by influencing the production of progesterone and estrogen.

The production of the hormones that control the menstrual cycle (progesterone, estrogen, FSH and LH) are regulated by feedback mechanisms.

To understand the menstrual cycle, we need to understand the structure of the female reproductive system (figure 37), including the ovaries (figure 38).

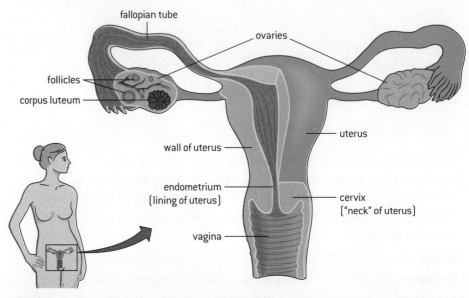

Figure 37. The female reproductive system

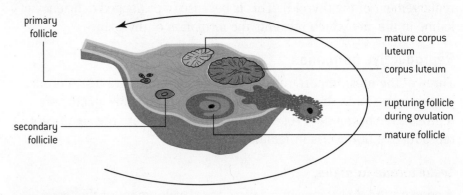

Figure 38. The structure of the ovary

The main parts of the female reproductive system include:

- The vagina—a muscular tube that connects to the cervix of the uterus. Its function is to receive sperm from the penis during sexual intercourse and transfer them to the uterus and the fallopian tubes. It also acts as a birth canal during child birth.

- The cervix—also called the neck of the uterus. It is a small muscular tube that connects the vagina to the uterus.

- The uterus—a hollow muscular organ that is connected to two fallopian tubes, one on each side. Its main function is to support the developing fetus. The lining of the uterus is called endometrium and its main function is to provide support for the fetus once implanted.

- Fallopian tubes—two muscular tubes, each connects the uterus to one ovary on each side. The inside of the fallopian tubes has cilia so that the ovum (egg) is easily transferred to the uterus.

- The ovaries—the site of *ovulation.* There are two ovaries located on the right and left side of the body. Each ovary acts as a small endocrine gland that produces female sex hormones: estrogen and progesterone. Females are born with a number of *follicles* inside the ovaries. The follicles at this stage are considered immature and they are called primary follicles. The follicle is like a sac that contains the immature egg which is also called *oocyte.* After puberty, during every menstrual cycle several follicles develop but only one ovulates (produces an egg). After ovulation, the follicle changes to become a *corpus luteum* that will usually disintegrate if fertilization does not take place.

Key term

Ovulation is the process by which an egg is released from the follicle inside the ovary.

The **follicle** is a sac-like structure that contains the egg.

The **oocyte** is the egg inside the immature follicle.

The menstrual cycle

The menstrual cycle starts at puberty and takes around 28 days.

At the beginning of the cycle, the pituitary gland produces FSH which stimulates the follicle to develop. As the follicle develops, the egg inside becomes mature. The follicle cells start releasing estrogen. Estrogen causes more secretion of the FSH, which in turn causes the estrogen level to increase. This is an example of **positive feedback** as FSH and estrogen have a complimentary relationship. Estrogen stimulates the lining of the uterus to thicken.

Towards the middle of the cycle, the high level of estrogen stimulates the pituitary gland to produce more LH. LH stimulates ovulation; the release of the egg from the follicle. After ovulation takes place, LH stimulates the follicle to turn into a corpus luteum. The corpus luteum secretes low levels of estrogen and high levels of progesterone. Progesterone maintains the lining of the uterus and prepares it for implantation (of the fetus).

By the end of the cycle, if fertilization does not take place, the corpus luteum disintegrates. This results in a drop in the amount of estrogen and progesterone. As a result, the lining of the uterus breaks down and menstruation takes place. The drop in the level of the reproductive hormones (estrogen and progesterone) stimulates the pituitary gland to produce more FSH and LH (negative feedback). FSH levels rise once again and a new menstrual cycle begins.

Table 5 summarizes the production and functions of the female reproductive hormones.

Table 5. A summary of the female reproductive hormones

Gland	Hormone	Function
Pituitary gland	Follicle-stimulating hormone (FSH)	• Stimulates the development of the follicle and the egg inside it • Stimulates the secretion of estrogen
	Luteinizing hormone (LH)	• Causes ovulation (egg release) • Stimulates the secretion of progesterone from the corpus luteum
Ovaries	Estrogen	• Stimulates the development of the endometrium (lining of the uterus) • Stimulates the pituitary gland to secret high levels of LH • Inhibits the secretion of FSH and LH
	Progesterone	• Maintains the lining of the endometrium to prepare it for implementation • Inhibits the secretion of FSH and LH

Question

Q

12 State four examples of hormones and their target tissue, and
explain the resulting effect.

Chapter summary

In this chapter, you have learned about the systems of the human body and their functions. Make
sure that you have a working knowledge of the following concepts and definitions:

☐ Digestion is the breakdown of large insoluble food molecules into small and soluble molecules so
that they can be absorbed into the blood stream.

☐ Digestive enzymes are needed for the digestion process. Digestive enzymes are carbohydrases,
proteases and lipases.

☐ Villi in the small intestine are the main site for absorption.

☐ The circulatory system is made of the heart, blood and blood vessels. The function of the heart is
to pump blood throughout the body.

☐ Blood vessels include veins, arteries and capillaries.

☐ Coronary arteries provide the heart with the oxygen and nutrients it needs.

☐ The function of the respiratory system is to bring in oxygen which is needed for aerobic respiration
and get rid of carbon dioxide which is a waste product.

☐ As we inhale or exhale, our ribcage and chest move to force air in and out.

☐ Cellular respiration is the process by which energy is released from organic compounds such as
glucose.

☐ The body has three lines of defence that protect it against pathogens which may cause diseases.

☐ Viruses are non-living organisms that reproduce inside living host cells.

☐ The nervous system consists of the CNS which consists of the brain and spinal cord, and the PNS
which consists of the nerve cells that connect all the parts of the body with the CNS.

☐ Nerve impulses are passed from one neuron to another.

☐ Hormones are chemical substances secreted from glands into the blood stream to regulate body
functions.

Additional questions

1. Explain the adaptation of the villi in the small intestine.
2. Describe the mechanism of ventilation in the human lung.
3. Explain the process of phagocytosis.
4. Explain what is meant by an antagonistic muscle pair.
5. Outline the structure of a nerve cell.
6. Explain how the nerve impulse is passed from one neuron to another.
7. Explain how goitre can develop in humans.
8. Explain the process of yeast fermentation.
9. Outline the stages of the menstrual cycle.

> I observed that plants not only have a faculty to correct bad air in six to ten days, by growing in it... but that they perform this important office in a complete manner in a few hours; that this wonderful operation is by no means owing to the vegetation of the plant, but to the influence of light of the Sun upon the plant.

Jan Ingenhousz, 1779

Chapter context

Plants are multicellular eukaryotes which make the base of most food webs. They have the ability to make their own food using water, carbon dioxide and light through a process called **photosynthesis**. Plants have two transport systems: **xylem** which transports water and **phloem** which transports food from leaves to the rest of the plant.

Learning objectives

In this chapter, you will learn about:

→ the process of **photosynthesis** and factors which may affect it

→ how water and food are transported inside plants.

🔑 Key terms introduced

→ Photosynthesis
→ Chlorophyll
→ Photolysis
→ Optimal level and limiting factor
→ Plateau
→ Xylem, phloem and vascular bundles
→ Transpiration
→ Active translocation

4.1 Photosynthesis

Photosynthesis is the process by which plants use inorganic water and carbon dioxide to produce organic compounds in cells using light energy. This process takes place in the chloroplasts of leaves.

Leaves

The leaf is the site of photosynthesis. To understand how photosynthesis takes place, we need to understand the structure of the leaf and how the leaf is adapted to perform its function. If we look inside the leaf, we will see the structures shown in figure 1.

🔑 Key term

Photosynthesis is the process by which plants use light energy, water and carbon dioxide to produce organic compounds.

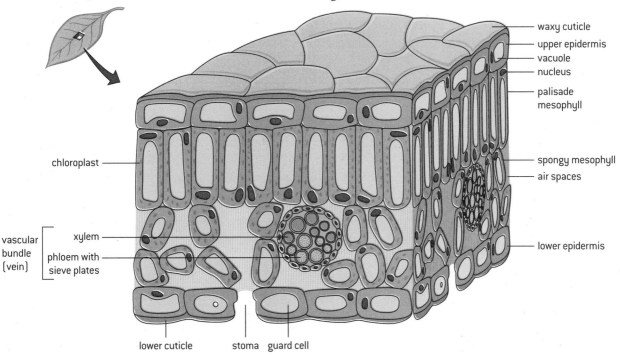

Figure 1. A cross-section of a leaf

Labels: waxy cuticle, upper epidermis, vacuole, nucleus, palisade mesophyll, spongy mesophyll, air spaces, lower epidermis, chloroplast, xylem, phloem with sieve plates, vascular bundle (vein), lower cuticle, stoma, guard cell

The structure of the leaf is adapted to enable it to carry out its functions, as described in table 1.

Table 1. Structure of leaves

Structure	Adaptation	Function
Cuticle	Made of wax	Prevents water loss
Upper epidermis	Thin and transparent	Allows light to pass through
Palisade mesophyll	Contains many chloroplasts	Absorbs light for photosynthesis
Spongy mesophyll	Has air spaces	Facilitates gas exchange; the air spaces allow rapid diffusion of oxygen and carbon dioxide
Vascular bundle (vein)	Consist of xylem and phloem	The xylem transports water, and the phloem transports the products of photosynthesis to the rest of the plant
Lower epidermis	Contain stomata	Allows carbon dioxide to diffuse into the leaf and oxygen and water vapour to diffuse out

Question

1 Explain how the structure of the leaf allows it to carry out photosynthesis.

DP link

The origin of eukaryotic cells and the endosymbiotic theory will be studied in **1.5 The origin of cells** in the IB Biology Diploma Programme.

Chloroplasts

Chloroplasts are tiny organelles in plant cells where photosynthesis takes place (figure 2 and table 2). They are mainly found in the palisade layer inside the leaf. Chloroplasts contain *chlorophyll*, which is the main pigment of photosynthesis. It is believed that chloroplasts were once photosynthetic bacteria that were ingested by larger prokaryotes to provide organic matter to the larger cell. The photosynthetic bacteria evolved over time to become chloroplasts. This theory is called the endosymbiotic theory, which explains the origin of eukaryotic cells.

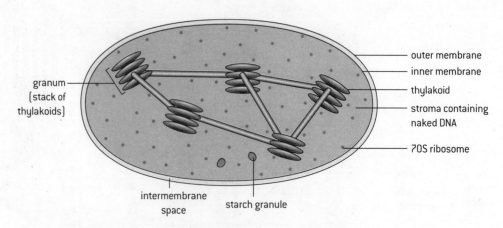

Figure 2. The structure of the chloroplast

Table 2. The main structure of the chloroplast

Structure	Adaptation
Stroma	Contains many enzymes which are important for photosynthesis.
Thylakoids	The site of photosynthesis. It contains chlorophyll which is needed for light absorption. Thylakoids have a large surface area to allow for more light absorption. Thylakoids are packed together in a structure called a granum.

Chlorophyll

Chlorophyll is the main photosynthetic pigment. Chlorophyll absorbs specific wavelengths of visible light (figure 3). It absorbs red and blue light most effectively, while it reflects green light; this is why it appears green in colour. There are different types of chlorophyll, mainly chlorophyll a and b, which differ very slightly in the wavelength they absorb most efficiently.

Key term

Chlorophyll is the main photosynthetic pigment which absorbs red and blue light most effectively and reflects green light.

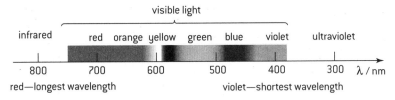

Figure 3. The range of wavelengths in visible light

The graph in figure 4 shows the percentage of light absorbed by chlorophyll for each wavelength of light. The following can be noted from figure 4:

- The highest absorption is seen with the violet–blue light. There is also good absorption with the red–orange light. This is because chlorophyll absorbs red and blue light most effectively.
- The lowest absorption is seen with the green–yellow light. This is because green light is reflected by chlorophyll, and therefore not absorbed.

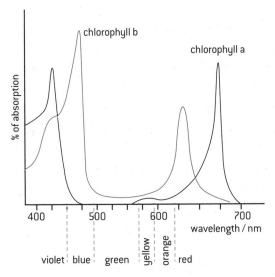

Figure 4. The absorption spectrum of chlorophyll a and b

The action spectrum (figure 5) is a graph that shows the rate of photosynthesis for each wavelength of light. Compare the action spectrum with the absorption spectrum of chlorophyll a and b (figure 4). It can be noted that the rate of photosynthesis is the least (at a minimum) at the green wavelength of light, while it is the highest with the blue and red light.

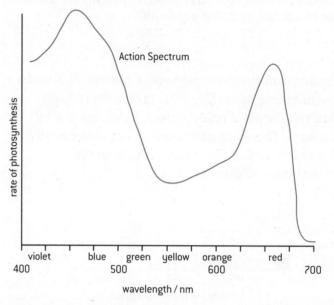

Figure 5. The action spectrum

Question

2   Explain why the green region in the action spectrum gives the lowest rate of photosynthesis.

Practical skills: Separation of photosynthetic pigments by chromatograph

Paper chromatography can be used to separate photosynthetic pigments. Plants and other photosynthetic organisms can have a combination of photosynthetic pigments such as chlorophyll, carotenes and others. The process of chromatography separates molecules based on the different solubility of these molecules in a specific solvent. You might carry out the following experiment in a DP Biology class (figure 6):

* Draw a horizontal line with a pencil about 1 cm from the bottom of a piece of chromatography paper.

* Use a small coin to press down a leaf on the line until you form a green line on the chromatography paper. Repeat this step until the line is fairly dark.

* Place the paper in a solvent (for example, ethanol) but make sure that the green spot is not touching the solvent.

* Allow it to stand for 20 to 30 minutes.

* Calculate the R_f value for each component observed on the filter paper:

$$R_f = \frac{\text{distance travelled by compound}}{\text{distance travelled by solvent}}$$

The more soluble the pigment, the further the movement of the pigment, and therefore the higher the R_f value. Different pigments may be identified by comparing their R_f values.

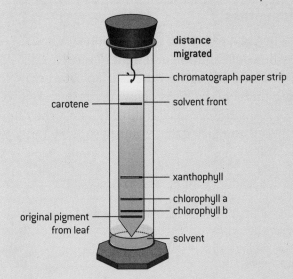

Figure 6. Separation of photosynthetic pigments by chromatograph

The process of photosynthesis

Photosynthesis is the process by which plants make use of inorganic water and carbon dioxide to produce organic compounds (carbon compounds) in cells using light energy.

$$\text{carbon dioxide} + \text{water} \xrightarrow[\text{absorbed by chloropyll}]{\text{Light energy}} \text{glucose} + \text{oxygen}$$

$$6CO_2 + 6H_2O \longrightarrow C_6H_{12}O_6 + 6O_2$$

Photosynthesis involves the conversion of light energy into chemical energy.

$$\text{Light energy} \xrightarrow{\text{Converted to}} \text{chemical energy (in glucose)}$$

There are two reactions that take place during photosynthesis, the light-dependent reaction and the light-independent reaction.

Stage 1: Light-dependent reaction

During the light-dependent reaction, light energy is absorbed by chlorophyll and is used for two main functions:

* To produce ATP, which is needed for the light-independent reaction
* *Photolysis* of water, where water molecules split to form hydrogen ions (H^+) and oxygen. Oxygen is lost as a waste product and hydrogen ions are needed for the light-independent reaction.

Stage 2: Light-independent reaction

During the light-independent reaction, energy (in the form of ATP) and hydrogen ions produced in the light-dependent reaction are used to produce carbohydrates from carbon dioxide.

$$\text{carbon dioxide} + \text{hydrogen ions} \xrightarrow{\text{ATP}} \text{glucose}$$

 Key term

Photolysis is the splitting of water molecules by light to form hydrogen ions (H^+) and oxygen.

DP link

Changes to Earth's
atmosphere, oceans
and rock deposition
due to photosynthesis
will be studied in **2.9
Photosynthesis** in the IB
Biology Diploma Programme.

Key term

The **optimal level** is where
the maximum rate of
photosynthesis is reached.

A **limiting factor** is any factor
that can limit the rate of
photosynthesis.

Question

3   What are the products of the light-dependent reaction?

Factors affecting photosynthesis

There are three factors which can affect the rate of photosynthesis:
temperature, carbon dioxide concentration and light intensity. Each
of these factors can limit the rate of photosynthesis if any are below
their *optimal level*. If more than one factor is below their optimal level,
only one factor will act as a limiting factor and directly affect the rate
of photosynthesis. Usually it is the factor that is furthest away from its
optimal level.

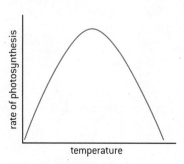

Figure 7. Effect of temperature on the rate of photosynthesis

At low temperatures the rate of photosynthesis is very low. As the
temperature increases, the rate of photosynthesis increases until the
maximum rate of reaction is reached at the 'optimum' temperature.
Above the optimum temperature the rate of photosynthesis starts
to decrease very rapidly as enzymes lose their shape and denature
(figure 7).

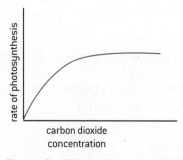

Figure 8. Effect of carbon dioxide concentration on the rate of photosynthesis

Key term

Plateau is the state of
minimal or no change in
a variable when another
variable is altered.

CO_2 is needed for the formation of carbon compounds in the light-
independent reaction. At low CO_2, the rate of photosynthesis is low. As
the CO_2 concentration increases, the rate of photosynthesis increases.
At high levels of CO_2, there is no further increase in photosynthesis
and the rate reaches a *plateau* (figure 8).

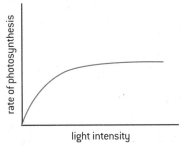

Figure 9. Effect of light intensity on the rate of photosynthesis

Light is used to produce ATP and split water by photolysis to form H^+ ions and oxygen. At low light intensity, the rate of photosynthesis is low. As light intensity increases the rate of photosynthesis increases until a certain point where further increase in light intensity will no longer increase the rate of photosynthesis. The rate of photosynthesis levels off and becomes constant (figure 9).

Question

4   Explain the effect of temperature on the rate of photosynthesis.

Maths skills: Measuring the rate of reaction

You can measure the rate of any reaction in two ways.

1. **Measuring the rate at which a product is formed.** For example, to measure the rate of photosynthesis, you can measure the rate at which oxygen is produced. This can be done in several ways, such as:

   • Count bubbles of oxygen produced from a pondweed.

   • Collect oxygen produced and measure its volume.

   • Use an oxygen probe to find the oxygen concentration.

2. **Measuring the rate at which a reactant is used up.** For example, to measure the rate of photosynthesis, you can measure the rate at which carbon dioxide is being used up. This can be done in several ways:

   • When carbon dioxide is absorbed from water, the pH of the water rises and this can be measured with pH indicators or pH meters.

   • Use a carbon dioxide probe to find the carbon dioxide concentration.

Once the amount of product or reactant is measured in a specific time, the rate can be calculated using one of the two methods:

a) $\text{Rate} = \dfrac{\text{amount* of product}}{\text{time}}$

*where amount refers to volume, concentration or number of bubbles.

b) from a graph (figure 10):

$$\text{Rate} = \text{slope} = \frac{\text{change in amount of product (change in } y)}{\text{change in time (change in } x)}$$

Worked example: Calculating the rate of reaction

WE

1. Calculate the rate of reaction using the graph in figure 10.

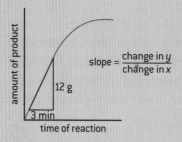

Figure 10. Calculating the rate of reaction

Solution

Use the formula below to find the value of the slope:

$$\text{Rate of reaction} = \text{Slope} = \frac{\text{change in } y}{\text{change in } x}$$

$$\text{Rate of reaction} = \text{Slope} = \frac{12 \text{ g}}{3 \text{ min}} = 4 \text{ g min}^{-1}$$

2. Calculate the rate of photosynthesis knowing that 10 cm³ of oxygen gas was collected in a syringe in 4 minutes.

Solution

Use the formula below to calculate rate of photosynthesis:

$$\text{Rate of photosynthesis} = \frac{\text{volume of O}_2 \text{ produced}}{\text{time}}$$

$$\text{Rate of photosynthesis} = \frac{10 \text{ cm}^3}{4 \text{ min}} = 2.5 \text{ cm}^3 \text{ min}^{-1}$$

Question

5   Calculate the rate of photosynthesis knowing that 15 oxygen gas bubbles were collected in three minutes.

Practical skills: Investigating the limiting factors of photosynthesis

You can use the apparatus in figure 11 to measure the effect of temperature on the rate of photosynthesis. You can place this apparatus in water baths of varying different temperatures, and calculate the rate of reaction by measuring the volume of oxygen gas produced in a period of time. Keep in mind the control variables that must be kept constant such as the light intensity in the room, carbon dioxide concentration and the volume of water used. Remember that you can investigate one factor at one time.

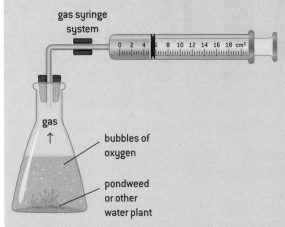

Figure 11. Measuring the effect of temperature on the rate of photosynthesis

DP ready   Nature of science

When designing an experiment it is essential to control the relevant variables. Consider the following variables (and how they relate to photosynthesis experiments such as the one shown in figure 11):

- Independent variable—this is the variable that you change. In figure 11, it is the temperature of the water bath.
- Dependent variable—this is the variable that you measure. In figure 11, it is the volume of oxygen gas collected in a period of time.
- Control variables—these are the variables that you must control so they do not impact the result of your experiment. In figure 11 these include light intensity, carbon dioxide concentration and volume of water.

Remember that you need to identify which limiting factor you plan to investigate—this will be your independent variable. The other limiting factors must be controlled and kept constant.

Key term

Xylem transports water from the roots to the leaves of a plant.

Phloem transports organic compounds (created by photosynthesis) from the leaves to the rest of the plant.

Vascular bundles are a collection of tube-like tissues including **xylem** and **phloem** which transport essential substances to the different parts of the plant.

4.2 Plant transport

Plants have two transport systems, one uses *xylem* which transports water from the roots to the leaves, and the second uses *phloem* which transports food from the leaves to the rest of the plant. Xylem and phloem form *vascular bundles* in the plant (figure 12).

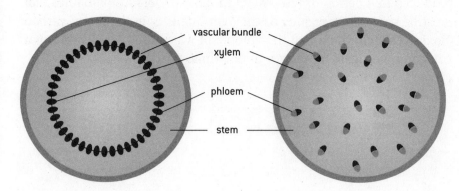

Figure 12. Vascular bundles in the stems of different types of plants

DP link

The transport in the xylem and phloem is discussed in **9.1 The transport in xylem** and **9.2 The transport in phloem (AHL)** in the IB Biology Diploma Programme.

Transport in xylem

Xylem vessels transport water from the roots to the leaves. Xylem tubes are made from dead cells which are strengthened by lignin to withstand the pressure from water. The diameter of the xylem vessels is wide to transport large amounts of water throughout the plant.

How is water transported through the xylem vessels?

- Water is absorbed from the soil into the roots. The root system is branched and the individual roots have root hairs (figure 13), thus increasing the surface area for water uptake.

- Water is absorbed by osmosis because the solute concentration inside the root is higher than in the soil, due to active transport of mineral ions into the root.

- Water is transported from one root cell to another until it reaches the xylem.

- Water is transported through the xylem vessels up the stem and then to the leaves.

- Water vapour evaporates from the spongy mesophyll cells and is lost via the stomata. Evaporation of water vapour via the stomata is called *transpiration*.

- Water lost is replaced from the xylem.

- This causes water from the roots to be pulled upward through the xylem. This pulling of water from the roots to the leaves against the force of gravity is called the transpiration stream.

Key term

Transpiration is the loss of water vapour from the leaves via the stomata.

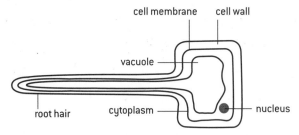

Figure 13. Root hair cell

The unique properties of water allow water to be pulled upward through the xylem because:

- Water molecules are cohesive; which means they can stick together. This is due to hydrogen bonding. This allows water to be drawn up the xylem tube.
- Water molecules are adhesive; which means they adhere to the wall of the xylem. This is due to their polarity. This helps to keep water within the xylem.

Factors that affect transpiration

There are four factors that affect the rate of transpiration: light, humidity, temperature and wind.

Light causes the stomata to open and therefore increases the rate of evaporation, which increases the rate of photosynthesis. However, in very dry conditions the stomata may close to reduce transpiration and water loss.

Humidity describes the concentration of water vapour in the atmosphere. When conditions are humid, there is a higher concentration of water vapour in the air than in the leaf. This causes a decrease in the rate of evaporation of water vapour from the leaf and therefore a decrease in transpiration rate.

As temperature rises, the transpiration rate increases. This is because increased temperature increases the rate of evaporation and the diffusion of water vapour from the leaf to the air outside.

Wind increases the transpiration rate. Wind blows away water vapour around the leaf, allowing for more effective diffusion of water vapour out of the leaf.

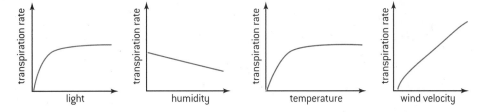

Figure 14. Factors affecting rate of transpiration

Question

6 Explain how some plants are adapted to reduce transpiration.
7 Suggest why the transpiration rate no longer increases when the temperature and light reach a certain point.

Practical skills: Measuring the rate of transpiration

A potometer is used to measure the rate of transpiration.

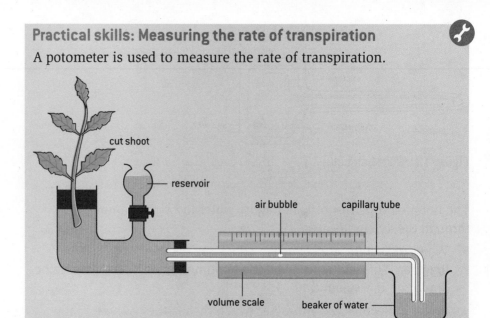

Figure 15. Potometer

The potometer measures the amount of water lost via transpiration. As water evaporates from the leaves, water moves up through the plant, causing the air bubble to move along the scale. This gives a measure of the volume of water absorbed by the plant (figure 15). Transpiration rate is the volume of water absorbed by the plant over a period of time.

Transport in phloem

In contrast to xylem, phloem consists of living tissue. The phloem tissue is made up of sieve tubes and companion cells (figure 16).

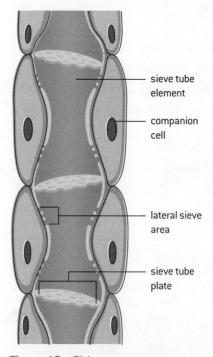

Figure 16. Phloem structure

- The companion cells are involved in ATP production.
- The sieve tubes are narrow elongated cells joined end-to-end, however, they no longer have a nucleus, so each sieve cell receives its instructions from the nucleus of the companion cell.

Phloem tubes transport organic compounds from photosynthetic tissues (leaves and stems) and storage organs to fruits, seeds and roots of the plant. Active transport is used to move organic compounds from photosynthetic tissue or storage organs into phloem sieve tubes. This transport is known as *active translocation* and requires ATP.

> **Key term**
>
> **Active translocation** is the process by which organic compounds in photosynthetic tissue or storage organs are moved into phloem sieve tubes. This process requires ATP.

Question

8 Outline the process of active translocation.

Practical skills: Using the *t*-test

You can use the *t*-test to compare between two sets of data. The *t*-test is a statistical test used to determine if your results are significant (or not) by comparing the mean values of two sets of data. We usually use the *t*-test when we have a large sample size (more than 10).

By running the *t*-test, you obtain the P value. The P value allows you to determine whether to accept or reject the null hypothesis. The P value indicates the probability that chance alone could produce the difference between the two sets of data.

For example, we may want to test if there is a significant difference between the mass of leaves growing in shade and the mass of leaves growing in light. We can come up with the following hypotheses:

Null hypothesis: there is no significant difference between the mass of leaves growing in shade and light.

OR

Alternative hypothesis: there is a significant difference between the mass of leaves growing in shade and light.

So, the question is: is there a significant difference between the mass of leaves growing in shade and light?

To answer that question, we need to calculate the P value and accordingly determine whether to accept or reject the null hypothesis. We can calculate the P value by processing the sets of data in Microsoft Excel.

What to conclude from the P value

If the P value < 0.05 ($< 5\%$)

- This means that probability that chance alone could produce the difference is $< 5\%$
- Which means that confidence level $> 95\%$ (confident that the difference is a real difference and not due to chance only)
- We can conclude that the difference is significant
- So, we reject the null hypothesis and accept the alternative hypothesis. This means that there is a significant difference between the mass of leaves growing in shade and growing in light.

If the P value > 0.05 ($> 5\%$)

- This means that probability that chance alone could produce the difference $> 5\%$
- Which means that confidence level $< 95\%$ (confident that the difference is due to chance only)
- We can conclude that there is no significant difference
- So, we accept the null hypothesis. This means that there is no significant difference between the mass of leaves growing in shade and growing in light.

Chapter summary

In this chapter, you have learned about the process of photosynthesis, the limiting factors which may affect it, and how water and food are transported inside plants. Make sure that you have a working knowledge of the following concepts and definitions:

- ☐ Plants are multicellular eukaryotes that make their own food.
- ☐ Photosynthesis is the process by which plants use inorganic water and carbon dioxide to produce organic compounds in cells using light energy.
- ☐ Photosynthesis takes place in leaves in the chloroplast.
- ☐ The structure of the leaf is adapted to allow it to carry out photosynthesis.
- ☐ Chlorophyll is the main photosynthetic pigment which absorbs red and blue light most effectively and reflects green light.
- ☐ Photosynthesis involves the conversion of light energy into chemical energy.
- ☐ There are two reactions that take place during photosynthesis. Light-dependent reactions and light-independent reactions.
- ☐ There are three factors which can limit the rate of photosynthesis: light intensity, carbon dioxide concentration and temperature. If any of these factors are below their optimal level, they can limit the rate of photosynthesis.
- ☐ Plants have two transport systems, the xylem which transports water from the roots to the leaves, and the phloem which transports food from the leaves to the rest of the plant.
- ☐ Xylem vessels transport water from the roots to the leaves. Xylem tubes are made from dead cells which are strengthened by lignin to withstand the pressure from water.
- ☐ Transpiration is the loss of water vapour from the leaves via stomata.
- ☐ There are four factors that affect the rate of transpiration: light, humidity, temperature and wind.
- ☐ Phloem tubes transport organic compounds from photosynthetic tissues (leaves and stems) and storage organs to fruits, seeds and roots of the plant by active translocation.

Additional questions

1. Outline the effect of light intensity on the rate of photosynthesis.
2. Explain how to measure the rate of photosynthesis.
3. Explain the role of water in photosynthesis.
4. State the source of the oxygen produced as a by-product of photosynthesis.
5. Explain how the root system is adapted to increase absorption of water uptake from the soil.
6. Describe how water is carried up the stem and into the leaves.
7. List three factors that affect the rate of transpiration.
8. Explain how wind affects the rate of transpiration from a leaf.

Genetics

"Increased knowledge of heredity means increased power of control over the living thing, and as we come to understand more and more the architecture of the plant or animal we realize what can and what cannot be done towards modification or improvement. "

Reginald Punnett, 1911

Chapter context

Genetics is the study of how traits are passed from parents to offspring. **Chromosomes** inside the nucleus of a cell are made up of many **genes** that control all our characteristics and are passed to offspring through the sex cells. **Haploid** sex cells are produced by **meiosis**. Each parent provides us with half of our chromosomes. Genes exist in different forms called **alleles**.

Learning objectives

In this chapter you will learn about:

→ the material of inheritance carried on **chromosomes**

→ **meiosis** and production of **haploid** sex cells

→ inheritance through **monohybrid crosses** including **autosomal**, **sex linkage**, **codominance** and **multiple alleles**

→ some common **genetic disorders**

→ biotechnology such as **DNA profiling** and **gene transfer**.

🔑 Key terms introduced

→ Chromosomes
→ Karyograms and karyotypes
→ Gene mutation
→ Genomes
→ Fertilization and zygotes
→ Meiosis and cytokinesis
→ Self-pollination and cross-pollination
→ Dominant and recessive alleles
→ Genetic disorders
→ DNA profiling

5.1 The material of inheritance

The nucleus of a cell contains long thread-like structures called *chromosomes* (figure 1). Chromosomes are made up of many *genes* that control all our characteristics. A gene is a small section of the DNA that codes for a specific protein in your body. DNA carries the genetic code that determines the characteristics of a living organism.

🔑 Key term

Chromosomes are the long, thread-like structures in the cell that carry genes.

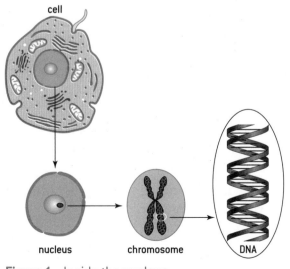

Figure 1. Inside the nucleus

Chromosomes

Chromosomes are long thread-like structures found in most living cells and carry genes which are responsible for the genetic information of the organism. The precise location of the chromosomes differs between prokaryotic and eukaryotic cells.

Prokaryotic chromosomes are made of circular DNA and are not enclosed in a nuclear membrane, whereas eukaryotic chromosomes are made of linear DNA and are found inside a nucleus. Prokaryotes have one chromosome while eukaryotes have many chromosomes.

The number of chromosomes is a characteristic feature of members of a species. For instance, humans have 46 chromosomes, chimpanzees have 48 chromosomes and dogs have 78 chromosomes.

> **DP ready** | **Theory of knowledge**
>
> In 1922, it was believed that the number of chromosomes in humans was 48. This idea remained established for 30 years even when photographic evidence showed that there were 46. What role, if any, does inertia play in the re-evaluation of existing beliefs?

Chromosomes usually exist as duplicates in preparation for cell division. A chromosome is made up of two chromatids which are joined by the centromere. These two chromatids in a chromosome are known as sister chromatids, which are the original chromatid and a copy (figure 2). The chromatids separate from each other during mitosis to form two new chromosomes.

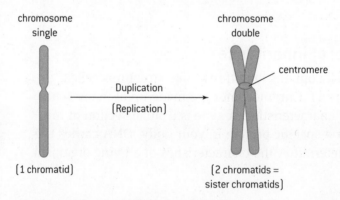

Figure 2. Structures of chromosomes

Chromosomes occur in pairs of *homologous chromosomes* in all body cells, except in sex cells. We can therefore define two types of cell: *diploid cells* and *haploid cells*.

- Diploid cells have two sets of chromosomes in the nucleus (one set from each parent) and these paired sets of chromosomes are called homologous chromosomes.

- Haploid cells have one set of chromosomes in the nucleus, which means they have one chromosome from each pair.

All body cells are diploid cells except for the sex cells (gametes) which are haploid cells.

In eukaryotes, there are different chromosomes that carry different genes:

- *Autosomes*, which are the chromosomes that do not determine sex. In humans, this includes 22 pairs of chromosomes.

- *Sex chromosomes*, which are the chromosomes that determine sex. In humans, this includes the last pair of chromosomes which is the 23rd pair. Females have two similar sex chromosomes (XX), while males have two different sex chromosomes (XY). The X and Y sex chromosomes are different in shape and size from each other, unlike the rest of the chromosomes (autosomes).

The autosomes and sex chromosomes of an organism can be arranged in homologous pairs, decreasing in length, to form an image called a *karyogram* (figure 3). A karyogram could be used for two main reasons:

- To deduce sex by looking at the last sex chromosome; if the two chromosomes in the last pair are similar in shape and size, then it is a female (XX), whereas if the two chromosomes in the last pair are different in shape and size, then it is a male (XY).

- To diagnose chromosomal abnormalities such as Down syndrome in humans.

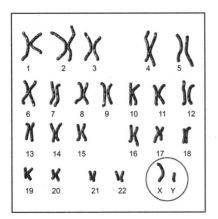 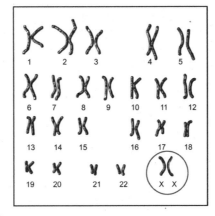

Figure 3. Left: male karyogram; right: female karyogram

The terms *karyotype* and karyogram have different meanings. Karyotype is a property of a cell which indicates the number and type of chromosomes present in the nucleus; it is not a photograph or diagram of them. The term karyogram is used to indicate the photograph of these chromosomes.

Karyotyping is important for pre-natal diagnosis because it can be used to identify a number of chromosomal abnormalities at an early stage. In order for pre-natal karyotyping to take place, cells are taken from the fetus and the fetal karyogram is analysed. Methods for obtaining fetal cells include:

- Amniocentesis—which involves the insertion of a needle into the amniotic fluid surrounding the developing baby and extracting some of the fluid where fetal cells are found.

- Chorionic villus sampling (CVS)—which involves taking a sample from the *chorionic villus* found in the placenta, where some fetal cells are found.

 Key term

Autosomes describe any of the chromosomes that do not determine the sex of the organism. The chromosomes in an autosomal chromosome pair are of a similar size and shape to each other and carry the same sequences of genes.

Sex chromosomes are the chromosomes that determine sex.

 Internal link

Down syndrome will be studied later in this chapter in **5.2 Reproduction and meiosis.**

 Key term

A **karyogram** is an image that shows the chromosomes of an organism in homologous pairs of decreasing length.

Key term

Karyotype is the property of the cell that describes the number and type of chromosomes present in the nucleus.

 Internal link

Examples of karyotypes and karyograms will be given in **5.2 Reproduction and meiosis.**

Risks associated with the pre-natal testing include miscarriage, bleeding, infection, cramping and harming the fetus. Parents must be made aware of the associated risks and possible consequences before making a decision whether or not to proceed with the pre-natal testing or not.

Genes

A gene is a small section of DNA that codes for a specific protein in your body and influences a specific characteristic. A gene occupies a specific location (known as a *locus*) on a chromosome (figure 4). The different forms of a gene are called *alleles*.

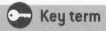
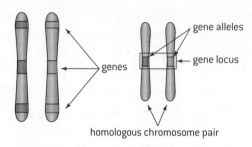

Figure 4. Location of genes on chromosomes

DP ready | **Nature of science**

Improved tools allow for new scientific discoveries

In the past, sequencing the entire human *genome* was an idea that seemed impossible. However, the improvement in technology towards the end of the 20th century led to the development of DNA sequencers. This made it possible to read the genome accurately and efficiently.

Figure 5. DNA sequencers at the University of Oxford

Gene mutation

A *gene mutation* is any change to the base sequence of a gene in an organism. Mutation can range from a change in a single nitrogen base to multiple nitrogen bases. There are many types of mutation including:

- Substitution—when a single base is replaced with another.
- Insertion—when an extra piece of DNA is inserted into the original DNA. This results in having additional bases in the gene.
- Deletion—when a part of the DNA is missing. This results in the absence of some bases.

Key term

A **gene mutation** is any change to the base sequence of a gene in an organism.

Figure 6. Types of gene mutation

A mutation may have no effect on the organism if it takes place in a region that does not change the amino acid sequence in the protein it codes for or if the mutation takes place in a non-coding region. If a mutation changes the amino acid sequence of the protein the gene codes for, it can cause serious issues in the organism. Sickle-cell anemia is an example of a substitution mutation that affects red blood cells in the body.

Mutations may be acquired during the organism's lifetime or may be passed on from parent to offspring. There are many factors that may cause mutations such as X-rays, UV radiation and viruses.

Internal link

Sickle-cell anemia will be studied later in this chapter in **5.3 Inheritance**.

DNA

DNA is a thread-like chain of nucleotides carrying the genetic instructions for the formation of proteins used in the growth and development of the organism.

The human genome is made of over 3 billion base pairs of DNA. Other organisms have different genome sizes.

DNA is made of coding and non-coding regions:

- The coding region codes for the formation of mRNA which is further translated to a sequence of amino acids that make up a protein. This means that the coding region codes for protein synthesis.
- The non-coding region does not code for mRNA formation and therefore is not involved in protein synthesis. Non-coding regions are referred to as junk DNA. However, it is believed that some of the non-coding regions of DNA code for the formation of other RNAs such as transfer RNA, which is needed for the DNA translation and protein synthesis. Other non-coding regions may act as genetic "switches", which determine where and when genes get expressed.

Internal link

The structure of DNA was studied in **2.5 Nucleic acids**.

DP link

DNA and RNA will be explained further in **2.6 Structure of DNA and RNA** in the IB Biology Diploma Programme.

Improved tools allow for new scientific discovery
From the 1940s onwards, autoradiography was used by biologists to reveal the structures found inside cells. John Carins used this technique and produced images of DNA molecules from *E. coli*. These images revealed that the chromosome in *E. coli* is made of a single circular DNA that is 100 µm in length. Other researchers then used this technique to reveal the length and shape of the DNA of different organisms.

5.2 Reproduction and meiosis

Sexual reproduction would not be possible without the process of meiosis. Meiosis is also responsible for the genetic variation observed in humans, animals and plants. In this section you will discover the role that meiosis plays in reproduction.

Reproduction

Reproduction is the process by which offspring are produced from their parents. There are two main ways of reproducing:

- Asexual reproduction which involves one parent. All offspring produced are genetically identical to the parent. This includes the binary fission of bacteria.

- Sexual reproduction involves the combining of male and female sex cells. A large amount of genetic variation is observed in offspring. Humans and most animals reproduce by sexual reproduction.

Fertilization

In males, the sex cell (or gamete) is called sperm, which is made inside the testes. In females, the sex cell is called the egg (or ovum), which is made inside the ovaries. Both sperm and eggs are haploid cells which means that, in humans, they have 23 chromosomes. During sexual reproduction, the sperm and egg fuse together to form the *zygote*, which has 46 chromosomes in total. This process is called *fertilization* (figure 7).

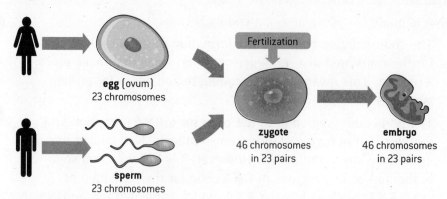

Figure 7. Fertilization

Gametes (sex cells) are produced inside sex organs (testes and ovaries) by a process called *meiosis*.

Key term

Fertilization is the process that occurs when the sperm and egg fuse to form a zygote.

A **zygote** is a fertilized egg.

Key term

Meiosis is the process that results in the formation of four haploid cells from a diploid cell. It involves two divisions: meiosis I and meiosis II.

Meiosis

Meiosis occurs in the sex organs: the testes and ovaries. It results in the formation of four haploid cells from a diploid cell; this means that the cells produced will have half the number of chromosomes which existed in the original cell. Meiosis is responsible for the genetic variation in sex cells produced. It involves two divisions classified as *meiosis I* and *meiosis II* (figure 9 on page 101).

Before meiosis takes place, cells pass through the interphase stage, in which DNA is replicated so that all chromosomes consist of two sister chromatids.

Meiosis I (1st division)

The phases of the first division in meiosis, meiosis I, are described in table 1. Meiosis I results in the production of two haploid daughter cells.

Table 1. Phases of meiosis I

Phase	Key points
Prophase I	• The replicated homologous chromosomes pair up with each other • The adjacent non-sister chromatids cross over • The replicated homologous chromosomes become visible under the microscope as they condense and coil • Spindle microtubules start forming at each pole of the cell. They begin to extend towards the equator of the cell • The nuclear membrane starts disappearing
Metaphase I	• The homologous chromosomes align at the equator of the cell • The spindle microtubules at each pole attach to the centromeres of the homologous chromosomes
Anaphase I	• The homologous chromosomes move apart from each other, each to an opposite pole • The sister chromatids of each chromosome are still connected by a centromere
Telophase I	• Each chromosome from the homologous pair is at an opposite pole of the cell • The nuclear membrane reforms • The spindle microtubules disappear • Chromosomes at each pole decondense and uncoil • *Cytokinesis* I then takes place, where the cell membrane divides to form two daughter haploid cells

 Key term

Cytokinesis is the division of the cytoplasm to split the cell into two daughter cells. Cytokinesis takes place in meiosis and mitosis.

What is crossing over?

Crossing over is the process by which homologous chromosomes pair up with each other and the adjacent non-sister chromatids exchange genetic material (figure 8). This results in genetic variation.

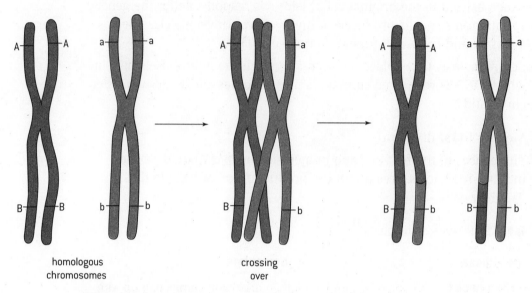

Figure 8. Crossing over

Meiosis II (2nd division)

The phases of the second division in meiosis, meiosis II, are described in table 2. Meiosis II results in the production of four haploid daughter cells.

Table 2. Phases of meiosis II

Phase	Key points
Prophase II	<ul><li>Each haploid cell produced by meiosis I passes through this stage</li><li>The chromosome at each pole is still composed of sister chromatids</li><li>Sister chromatids become visible under the microscope as they condense and coil</li><li>Spindle microtubules start forming at each pole of the cell</li><li>The nuclear membrane starts disappearing</li></ul>
Metaphase II	<ul><li>Sister chromatids align at the equator of the cell</li><li>The spindle microtubules at each pole attach to the centromeres of the sister chromatids</li></ul>
Anaphase II	<ul><li>The spindle microtubules shorten and pull the sister chromatids apart from each other, each to an opposite pole</li><li>This results in the splitting of the centromere connecting the sister chromatids</li><li>Each sister chromatid will then become a chromosome</li></ul>
Telophase II	<ul><li>Each chromosome is at an opposite pole of the cell</li><li>The nuclear membrane reforms around each group of chromosomes</li><li>The spindle microtubules disappear</li><li>Chromosomes at each pole decondense and uncoil</li><li>Cytokinesis II then takes place where the cytoplasm and cell membrane divide to form four daughter haploid cells</li></ul>

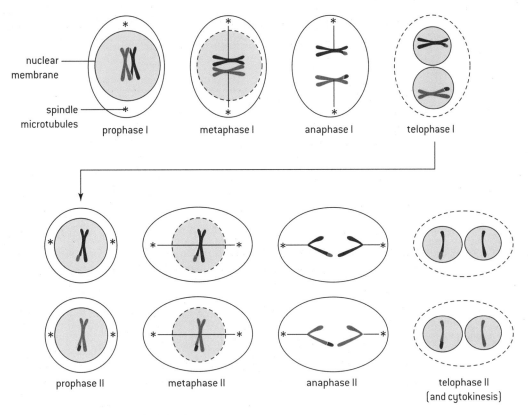

nuclear membrane

spindle microtubules

prophase I   metaphase I   anaphase I   telophase I

prophase II   metaphase II   anaphase II   telophase II (and cytokinesis)

Figure 9. The phases of meiosis

Question

1   Distinguish between meiosis and mitosis.
2   Describe the process in meiosis that promotes genetic variation.

Internal link

Mitosis was discussed in **1.5 Cell division**.

DP ready | Nature of science

Meiosis was discovered by microscope examination of dividing sex cells. The development of the advanced electronic microscopes in the 19th century allowed scientists to get detailed images of the nucleus and the chromosomes inside it. The use of these microscopes revealed that the nuclei of the sex cells (sperm and egg) of the horse threadworm contained two chromosomes while the fertilized egg contained four. This indicated that the sex cells had half the number of chromosomes found in the fertilized egg.

Note the following with regards to the chromosomes in meiosis I and II:

• At the beginning of meiosis I, refer to the chromosomes as homologous chromosomes.
• At the beginning of meiosis II, refer to the chromosomes as sister chromatids.

Worked example: Stages of meiosis

1. Which stage of meiosis is shown in figure 10?

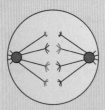

Figure 10. What stage of meiosis is this?

Solution

The chromosomes shown in figure 10 are separate from each other and moving in opposite directions, therefore, figure 10 represents the stage of anaphase. Since the chromosomes on each side are shown as sister chromatids, this indicates that the homologous pair has been split and each chromosome with its copy is moved to an opposite direction. This means that this stage is anaphase I. In anaphase II, we will not see sister chromatids on one side—you will notice that sister chromatids being split instead (and each copy is moved to opposite direction) as shown in figure 11.

Figure 11. Anaphase II

2. Look at figure 12 and answer the following questions:
   a) What stage of meiosis is shown?
   b) What is the number of chromosomes shown in this stage?
   c) What is the number of chromosomes in the parent cell?

Figure 12. Cell undergoing meiosis

Solution

a) The chromosomes shown in figure 12 are lined up at the equator of the cell and therefore it represents the stage of metaphase. Since the chromosomes are shown as sister chromatids, this indicates that the sister chromatids are about to split. This means that this stage is metaphase II. In metaphase I, we will not see sister chromatids at the middle but instead you will notice the homologous pair at the middle as shown in figure 13.

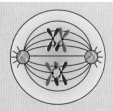

Figure 13. Metaphase I

b) There are two chromosomes, each existing as sister chromatids (the original and the copy).

c) Since this stage is metaphase II, the parent cell had the double of the number of chromosomes that are present. Note that the number of chromosomes is halved in meiosis I. This means that the parent cell had four chromosomes.

Question

3 Describe what happens to chromosomes during meiosis in a cell that contains four chromosomes.

Non-disjunction

Non-disjunction takes place during the formation of gametes. It happens either in anaphase I of meiosis I or anaphase II of meiosis II when the chromosomes fail to separate correctly. The gametes produced will either have an extra chromosome or a missing one. The sperm or egg which have a missing chromosome usually dies fast, whereas the sperm or egg with the extra chromosome usually survives. During the fertilization of these gametes, the zygote formed will have an extra chromosome.

An example of a genetic disorder resulting from non-disjunction is Down syndrome (trisomy 21), in which a person has a total of 47 chromosomes instead of 46 (figure 14). Studies have shown that the chance of offspring having Down syndrome increases with the age of the mother. Common symptoms of Down syndrome include delays in physical growth and intellectual disability that range from mild to severe.

🔑 **Key term**

Non-disjunction causes the production of gametes that either have missing or extra chromosomes. It occurs when the chromosomes do not separate properly during meiosis.

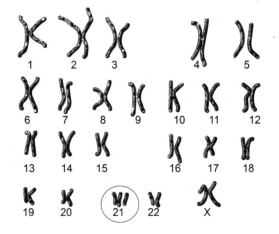

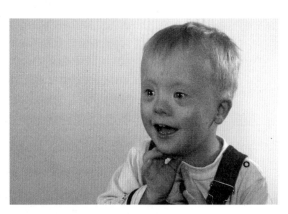

Figure 14. Left: Down syndrome karyogram; right: a child with Down syndrome

4 The *karyogram* in figure 15 shows the karyotype of a newborn.

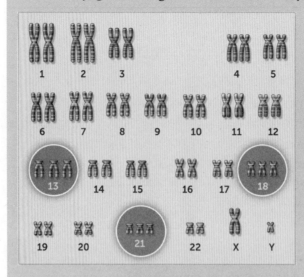

Figure 15. Karyotype of a newborn

a) Deduce with a reason the sex of the newborn.

b) Explain if the newborn shows any chromosomal abnormalities.

Worked example: Abnormalities in meiosis WE

3. Look at figure 16 and answer the following questions:

   a) Which stage of meiosis is shown?

   b) What abnormality can be noticed?

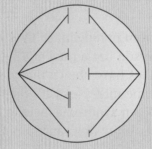

Figure 16. Cell during meiosis

Solution

a) The chromosomes shown in figure 16 are separate from each other and moving in opposite directions and therefore it represents the stage of anaphase. Since the chromosomes on each side are shown as one copy of the sister chromatids, this indicates that the sister chromatids have been split and each copy has moved to an opposite direction. This means that this stage is anaphase II.

b) It can be noticed that there was an unequal split of the sister chromatids which resulted with one side having an extra chromosome. When cytokinesis takes place, two gametes would be produced from this unequal split: a gamete with an extra chromosome and a gamete with a missing chromosome. This is an example of non-disjunction.

Practical skills: Identifying stages of meiosis using a microscope

You can observe meiotic cells under the microscope and identify the stages of meiosis I and II as chromosomes become visible. The main difference between meiosis I and II is the appearance of chromosomes, as shown in figure 17.

To distinguish meiosis I from meiosis II, note that the chromosomes in meiosis I will appear as XX, which indicates the replicated homologous pair of chromosomes. In meiosis II, chromosomes will appear as X, which indicates sister chromatids (the original and the copy).

To identify prophase, note how the chromosomes are visible, forming a cloud of very thin threads. If the chromosomes appear in pairs (XX), then it is prophase I. If the chromosomes appear as sister chromatids (X), then it is prophase II.

To identify metaphase, note how chromosomes are aligned at the equator of the cells. If chromosomes appear in pairs (XX), then it is metaphase I. If chromosomes appear as sister chromatids (X), then it is metaphase II.

To identify anaphase, note how the chromosomes are separating from each other. If the chromosomes on one side appear as one pair (X), then it is anaphase I. If the chromosomes appear as a single chromatid (I), then it is anaphase II.

To identify metaphase, note how the cell is elongated and in the process of dividing to two. If the chromosomes appear as sister chromatids (X), then it is telophase I. If the chromosomes appear as single chromatids (I), then it is telophase II.

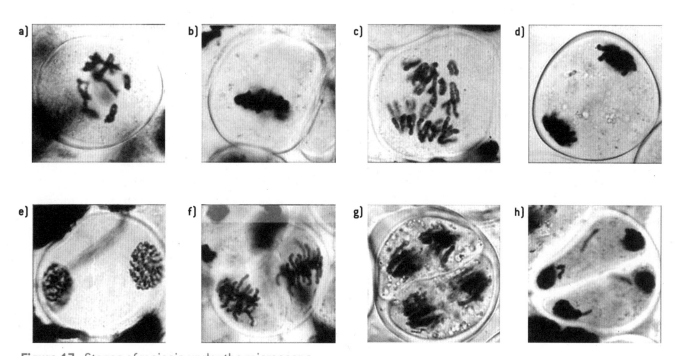

Figure 17. Stages of meiosis under the microscope:
(a) prophase I; (b) metaphase I; (c) anaphase I; (d) telophase I;
(e) prophase II; (f) metaphase II; (g) anaphase II; (h) telophase II

Figure 18. Gregor Mendel

5.3 Inheritance

Gregor Mendel (figure 18) was an Austrian monk and a teacher who discovered the basic principles of inheritance based on his experiments in his garden using pea plants. Mendel studied the characteristics in pea plants and how they were passed from one generation to another. He selected pea plants because it was easy to observe the traits of such plants. Mendel studied seven traits of pea plants (table 3). It was also easy for him to *self-pollinate* or *cross-pollinate* the flowers to produce offspring in a short period of time. He studied one trait at a time, collected data based on the traits observed in offspring produced and looked for patterns in his results. Based on his pea plant studies, Mendel proposed that traits are always controlled by single genes. According to Mendel, inheritance is based on passing such genes from parents to offspring. Mendel's work was not recognized until after his death. Gregor Mendel is now considered the father of genetics.

Table 3. The seven traits studied by Mendel

seed shape	seed color	pod shape	pod color	flower color	flower location	plant size
round	yellow	inflated	green	purple	axial	tall
wrinkled	green	constricted	yellow	white	terminal	short (dwarf)

Mendel made the following conclusions based on his studies:

- "Units" are passed on to offspring which determine the inheritance of each trait (these "units" are now called genes).
- An individual inherits one "unit" (gene) from each parent for each trait.
- A trait may not show up in an individual but can still be passed on to the next generation.

Unless you are able to repeat an experiment a number of times, you will not know if the results you obtained are reliable or not. Mendel crossed large numbers of pea plants, generating numerical data accordingly. He made sure he had replicates for his experiments to ensure the reliability of data. He used the repeats to see how close his results were. Mendel is considered a pioneer in obtaining numerical data (quantitative data) and following a scientific research method. It is now a standard in scientific research to have repeats to ensure reliability of results obtained.

The scientific community was slow to accept Mendel's theories. What factors would influence the scientific community to accept (or dismiss) new ideas? Mendel's work was neglected by scientists for a long period of time and many had doubts about his data. Many criticized that his data cannot be applied to most species. His data were rediscovered 20 years after his death. The rediscovery of his work was because of the focused public attention on heredity during that time. Many scientists conducted experiments based on Mendel's studies and reached similar conclusions to Mendel. Mendel's studies on pea plants provided the foundation principles of genetic inheritance.

Question

5   Suggest why Mendel chose pea plants to conduct his experiments.

The basics of genetics

In this section, we will focus on *monohybrid crosses* which refer to the study of the inheritance of a single trait.

The inheritance of traits is controlled by genes. A gene has different forms called alleles. The alleles that you have are referred to as *genotype* while your appearance or characteristic is referred to as *phenotype.*

We represent a gene by using letters as a symbol. For example, the colour of pea seeds is referred to as R or r.

If the allele affects the phenotype when present it is called a *dominant allele.* The dominant allele is represented by an uppercase letter. For example: yellow colour of pea seeds is a dominant trait, so we represent it as R.

 Key term

A **monohybrid cross** is the study of the inheritance of a single trait.

 Key term

Genotype describes the alleles of a gene carried by an organism.

Phenotype describes the expression of the gene; the characteristics of an organism.

Key term

A **homozygous** organism has two identical alleles of a gene.

A **heterozygous** organism has two different alleles of a gene.

A **dominant allele** is an allele that affects the phenotype when present in the homozygous or heterozygous state. Dominant alleles mask the effects of recessive alleles.

A **recessive allele** is an allele that affects the phenotype only when present in the homozygous state.

If the allele does not affect the phenotype when present with the dominant allele, and its effect is masked by the dominant allele, it is called a *recessive allele*. The recessive allele is represented by a lowercase letter. For example: green colour of pea seeds is a recessive trait, so we represent it by r.

If the genotype of an organism consists of two identical copies of an allele for a particular gene, then that organism is *homozygous* for that particular gene; for example, RR or rr.

If the genotype of an organism consists of two different alleles for a particular gene, then that organism is *heterozygous* for that particular gene; for example, Rr.

An example of Mendel's experiments

Mendel crossed a pea plant with yellow seeds with a pea plant with green seeds, and all the offspring were pea plants with yellow seeds. When these offspring were crossed with pea plants with green seeds, approximately one-half of the offspring were plants with green seeds.

Mendel explained his results as follows:

Since all offspring of the first cross had yellow seeds, then yellow must be dominant over green for the colour of seeds. All offspring had yellow seeds, this indicates that they all carry the dominant allele (R). We can construct a Punnett grid to predict the outcomes of genetic crosses.

Worked example: Constructing Punnett grids

4. In this worked example, the parent generation (F_0) of some yellow and green peas are crossed. A second cross between the yellow peas produced in the first cross (first generation offspring or F_1) and some green peas is also made. Using Punnett grids, find the genotypes and phenotypes of the offspring of each cross.

Solution

To make a Punnett grid, the gametes of both parents should be clearly labelled on each side of the grid. The four possible outcomes should be shown in the grid. The grid is useful in showing all the possibilities of the offspring genotypes and the overall ratio of each.

First cross

Key:

*   Dominant: yellow (R)
*   Recessive: green (r)

Parent generation (F_0) Phenotype: yellow peas × green peas

Genotype:	RR ×	rr
Gametes:	R	r

First generation (F_1)

	r	r
R	Rr	Rr
R	Rr	Rr

F_1 Genotype: all Rr

Phenotype: all yellow

Second cross

Phenotype: yellow peas (F₁) × green peas

Genotype:	Rr	×	rr

Genotype: Rr   ×   rr

Gametes: R, r   r

Second generation (F₂)

	r	r
R	Rr	Rr
r	rr	rr

F₂ Genotype: 50% Rr, 50% rr

Phenotype: 50% yellow, 50% green

Ratio: 1:1

Inheritance in humans

According to Mendel's studies on pea plants, he concluded that each of the seven traits he studied were controlled by a single gene. However, modern studies have revealed that most traits in humans are controlled by multiple genes as well as environmental factors such as ultraviolet (UV) light and exercise. Such studies have shown that inheritance in humans does not necessarily exhibit a simple Mendelian pattern of inheritance.

Some examples of dominant and recessive characteristics in humans are shown in table 4 on page 110.

DP link

Traits affected by multiple genes are called polygenic traits and will be studied in **10.2 Inheritance (AHL)** in the IB Biology Diploma Programme.

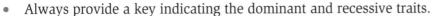

DP ready   Approaches to learning

When solving genetic problems, consider the following tips:

- Always provide a key indicating the dominant and recessive traits.
- Always choose a letter to represent a gene where the upper and lower cases are *clearly* different (R and r).
- Always set out an answer fully, clearly showing:
  - Parental generation (F₀)
  - Gametes
  - Offspring (F₁ or F₂).
- Construct a Punnett grid to show the offspring genotype possibilities.
- Notice that the phenotype is expressed in words whereas the genotype is expressed in letters.

Table 4. Examples of dominant and recessive alleles in humans

dominant gene		recessive gene	
cleft chin		no cleft	
widow's peak		no widow's peak	
dimples		no dimples	
freckles		no freckles	
free earlobe		attached earlobe	

Worked example: Inheritance

5. A mother and a father are both heterozygous with freckles.
Predict the phenotypes of their children.

Solution

Knowing that having freckles is dominant over not having freckles,
the following key is to be written:

- Dominant: freckles (F)
- Recessive: no freckles (f)

Since both parents are heterozygous with freckles, their genotype is Ff.

Parent generation (F_0)

Phenotype: mother with freckles (heterozygous) × father with
freckles (heterozygous)

Genotype: Ff × Ff

Gametes: F, f F, f

First generation (F_1)

	F	f
F	FF	Ff
f	Ff	ff

F_1 Genotype: FF, 2 Ff, ff

F_1 Phenotypes: FF—with freckles

Ff—with freckles

ff—without freckles

There is a 75% chance that their children will have freckles and a
25% chance they will not.

Question

6   Predict the phenotypes of the children of a mother who is heterozygous for having dimples and a father who has no dimples.

7   In pea plants, round seeds are dominant over wrinkled. A heterozygous plant for round seeds was crossed with a plant with wrinkled seeds. If there were 80 offspring produced, how many will have wrinkled seeds?

Test cross

If someone is showing a dominant trait, it could mean that the genotype is homozygous or heterozygous. For example, if someone has free ear lobes (a dominant trait), the genotype could be homozygous (EE) or heterozygous (Ee). So how do we know if someone who is showing a dominant trait is homozygous or heterozygous?

To determine whether a dominant phenotype is homozygous or heterozygous, we have to carry out a *test cross*, in which the unknown is crossed with a homozygous recessive.

- If the unknown is homozygous dominant, then 100% of offspring will show the dominant phenotype.

- If the unknown is heterozygous, then 50% of offspring will show the dominant phenotype and the other 50% will show recessive.

> **Key term**
>
> A **test cross** describes testing an unknown genotype by crossing it with a homozygous recessive.

Worked example: Test crosses WE

6.   A long-winged fly of an unknown genotype is mated in a test cross. Half of the offspring have long wings. What can be concluded from this result about the unknown genotype of the parent fly?

Solution

In a test cross, if 50% of offspring show the dominant phenotype (long-winged), this means that the unknown is heterozygous (Gg).

Question

8   A parent organism of an unknown genotype is mated in a test cross. All offspring have the same phenotype as the parent. What can be concluded from this result?

Codominant alleles

Mendel's studies were based on dominant and recessive alleles in his pea plants. However, some alleles are not dominant or recessive but are equally strong, and when present, they are both expressed in the phenotype. Such alleles are called *codominant alleles* (figure 19).

> **Key term**
>
> **Codominant alleles** are pairs of alleles which are both expressed when present. Codominant alleles have joint effects.

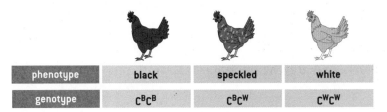

phenotype	black	speckled	white
genotype	$C^B C^B$	$C^B C^W$	$C^W C^W$

Figure 19. Codominance

Figure 20. Codominant alleles in cats can result in patterned tabby coats. Be careful this doesn't happen to your IB textbooks!

For example, a cross between a red flower plant with a white flower plant may result in a pink flower. In this case both alleles for red and white are codominant. You may represent codominance as follows, with symbols in the superscript to indicate the different characteristics:

- Red allele: C^R
- White allele: C^W

Parent generation (F_0)

Phenotype: red flower × white flower

Genotype: $C^R C^R \times C^W C^W$

Gametes: $C^R, C^R \quad C^W, C^W$

First generation (F_1)

	C^W	C^W
C^R	$C^R C^W$	$C^R C^W$
C^R	$C^R C^W$	$C^R C^W$

F_1 Genotype: $C^R C^W$

Phenotype: All pink.

Worked example: Codominance

7. A rooster with grey feathers is mated with a hen of the same phenotype. Among their offspring, 12 are grey, six are black and six are white.

   a) What is the explanation for the inheritance of these colours?

   b) What are the predicted genotypes for these chickens?

Solution

a) Since the cross between the grey rooster and the grey hen produced three different colours, then the genes for colour in chicken exhibit codominance as all were expressed and not masked by each other.

b) Since the alleles are codominant, they are presented as follows:

   - Black allele: C^B
   - White allele: C^W

Parent generation (F_0)

Phenotype: grey rooster × grey hen

Genotype: $C^B C^W \times C^B C^W$

Gametes: $C^B, C^W \quad C^B, C^W$

First generation (F_1)

	C^B	C^W
C^B	$C^B C^B$	$C^B C^W$
C^W	$C^B C^W$	$C^W C^W$

F_1 Genotype: $C^B C^B$, 2 $C^R C^W$, $C^W C^W$

Phenotype: Black, two grey, white (ratio = 1:2:1).

Question

9 In horses, the genes for colour exhibit codominance. The cross between a chestnut horse and a white horse produces a roan horse. What offspring would you expect from the mating of a roan horse and a white horse?

Inheritance of blood groups

Inheritance of ABO blood groups in humans is an example of codominance and *multiple alleles*. This inheritance is controlled by three alleles:

- I^A which is a dominant allele and relates to blood group A
- I^B which is a dominant allele and relates to blood group B
- i which is a recessive allele

> **Key term**
>
> **Multiple alleles** occur when some genes have more than two alleles.

Since I^A and I^B are dominant alleles, both are expressed when present together. This results in the formation of blood type AB. The allele (i) is recessive to both (I^A) and (I^B) and if present with any of the dominant alleles, its expression is masked. This means that if a person has the dominant allele (I^A) with the recessive allele (i), he will have blood type A. If a person has the dominant allele (I^B) with the recessive allele (i), he will have blood type B. Blood type O is only possible when both alleles are recessive (table 5).

Table 5. Possible genotypes for ABO blood groups

Phenotype	Genotype
A	$I^A I^A$ or I^Ai
B	$I^B I^B$ or I^Bi
AB	$I^A I^B$
O	ii

It is essential in blood transfusion to know the type of blood being transfused and the blood group of the patient. If a blood group is transfused where an antigen is introduced into the body, the body's immune response against such blood will be triggered which will result with complications due to the coagulation of red blood cells (table 6).

Table 6. Antigens in the different blood groups

	group A	group B	group AB	group O
red blood cell type	A	B	AB	O
antigens in red blood cell	A antigen	B antigen	A & B antigen	none

Worked example: Blood groups

8. If a man has blood group AB and a woman has blood group O, what is the probability that their child will be blood group O?

Solution

Parent generation (F_0)

Phenotype: a man with blood group AB × a woman with blood group O

Genotype: $I^A I^B$ × ii

Gametes: I^A, I^B × i, i

First generation (F_1)

	i	i
I^A	$I^A i$	$I^A i$
I^B	$I^B i$	$I^A i$

The probability that their child will be blood group O (ii) is 0%.

Question

Q

10 If a male with blood group A and a female with blood group B have a child with blood group O, what are the genotypes of the father and mother?

Sex-linked inheritance

Sex chromosomes include the X chromosome and the Y chromosome. Sex chromosomes determine gender:

- Females have two X chromosomes
- Males have one X chromosome and one Y chromosome.

Gender is determined by the chromosomes present in sperm and eggs. The chromosomes present in eggs are always X chromosomes. However, sperm may have either an X or Y chromosome. If the sperm has the Y chromosome, the forming zygote will develop into a male. If the sperm has the X chromosome, the forming zygote will develop into a female.

What are the chances of offspring being male or female?

This Punnett grid is used to predict the chance of having a boy or a girl.

		Father XY chromosomes	
		X	Y
Mother XX chromosomes	X	XX	XY
	X	XX	XY

Therefore, there is a 50% chance that offspring will be male (XY) and 50% chance that offspring will be female (XX).

The X chromosome is relatively large compared to the Y (figure 21). Some genes are present on the X chromosome and absent from the shorter Y chromosome in humans. If the gene located on the X chromosome (and absent from the Y chromosome) controls a specific characteristic, then this characteristic is associated with gender. This is called a *sex-linked trait* or *sex linkage*.

Since females have two X chromosomes they have two copies of the sex-linked gene whereas males only have one copy since they only have one X chromosome.

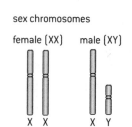

sex chromosomes

female (XX)   male (XY)

X X   X Y

Figure 21. Sex chromosomes

Red–green colour blindness

Red–green colour blindness is an example of a sex-linked recessive condition where the gene is carried on the X chromosome (table 7).

Table 7. Possible genotypes and phenotypes for red–green colour blindness

Genotype	Phenotype
$X^B Y$	Unaffected male
$X^b Y$	Affected male
$X^B X^B$	Unaffected female
$X^B X^b$	Unaffected female (carrier)
$X^b X^b$	Affected female

Key: X^B = unaffected allele; X^b = affected allele

> **Key term**
>
> A **sex-linked trait** or **sex linkage** occurs when a trait that is controlled by a gene is located on the sex chromosome and is associated with gender.

> **Key term**
>
> A **carrier** is an individual that has one copy of a recessive allele that causes a genetic disease. This allele does not affect the phenotype.

Worked example: Colour blindness   WE

9. What phenotypes would you expect for a cross between a female carrier for red–green colour blindness and a normal vision male?

Solution

A female carrier for red–green colour blindness will have a genotype $X^B X^b$

A normal vision male will have a genotype $X^B Y$

Parent generation (F_0)

Phenotype: an unaffected male × a female carrier for colour blindness

Genotype: $X^B X^b \times X^B Y$

Gametes: $X^B, X^b \times X^B, Y$

First generation (F_1)

	X^B	Y
X^B	$X^B X^B$	$X^B Y$
X^b	$X^B X^b$	$X^b Y$

The possible phenotypes are an unaffected female (50%), an unaffected male (25%) or an affected male (25%).

Question   Q

11 A woman with unaffected vision mates with a man, also with unaffected vision, producing a boy with red–green colour blindness. What are the genotypes of both parents?

Other inherited diseases

Genetic disorders are heritable diseases or conditions that occur as a result of a specific defect or mutation in a single gene.

Cystic fibrosis

Cystic fibrosis is an *autosomal* recessive disease that is caused by a mutation in the gene that codes for the formation of the chloride ion channels. This mutation causes a disruption in the function of chloride ion channels, affecting water movement across cell membranes. As a result, cells that line the lungs, pancreas and other organs produce mucus that is unusually thick and sticky. This mucus clogs the airways causing the characteristic signs and symptoms of cystic fibrosis.

Table 8. Possible genotypes and phenotypes for cystic fibrosis

Genotype	Phenotype
FF	Unaffected
Ff	Unaffected
ff	Affected

Key: F = Unaffected allele; f = affected allele

Worked example: Inherited diseases

10. Calculate the phenotypic ratio of a cross between a female who has cystic fibrosis with an unaffected man who is a carrier for cystic fibrosis.

Solution

Knowing that cystic fibrosis is an autosomal disease and that **not** having cystic fibrosis is dominant over having it, the following key is to be written:

- Dominant: Unaffected (F)
- Recessive: Affected (f)

The mother has cystic fibrosis, so her genotype is ff.

The father is unaffected but a carrier for cystic fibrosis, so his genotype is Ff.

Parent generation (F_0)

Phenotype: affected mother × unaffected father (but a carrier)

Genotype: ff × Ff

Gametes: f, f   F, f

First generation (F_1)

	F	f
f	Ff	ff
f	Ff	ff

F_1 Genotype: 2 Ff, 2 ff

F_1 Phenotypes: Ff—unaffected (but a carrier)

ff—Affected

There is a 50% chance of children having cystic fibrosis and a 50% chance they will not (ratio = 1:1).

Question

12 What offspring phenotypes would you expect for a cross between unaffected parents that are heterozygous for cystic fibrosis?

Sickle-cell anemia

Sickle-cell anemia is an autosomal codominant disease that affects the shape of red blood cells in the body. It is caused by a base mutation on the Hb gene, which codes for the formation of hemoglobin. This results in the formation of the abnormal hemoglobin (S) in red blood cells instead of the normal hemoglobin (A). As a result, sickle-shaped red blood cells are produced instead of the normal donut-shaped red blood cells. Sickle-shaped red blood cells cannot carry oxygen as efficiently as normal red blood cells would. Symptoms may include fatigue and shortness of breath. In addition, the sickled red blood cells may get stuck in the small blood capillaries causing pain.

The sickle-shaped red blood cells give resistance to malaria and, as such, sickle-cell anemia can be considered an advantage in areas of the world where malaria is found. However, individuals with homozygous (Hb^SHb^S) sickle-cell anemia are likely to have shorter lifespans, whereas individuals with heterozygous (Hb^SHb^A) sickle-cell anemia carry the trait **and** half of their red blood cells are sickled (table 9). These individuals are known as carriers or *heterozygotes*. Heterozygotes remain healthy with only half of their red blood cells affected.

Table 9. Possible genotypes and phenotypes for sickle cell anemia

Genotype	Phenotype	Relation to malaria
$Hb^A Hb^A$	Unaffected	Not resistant to malaria
$Hb^A Hb^S$	Half unaffected/ half affected	Malaria symptoms less severe
$Hb^S Hb^S$	Affected	Resistant to malaria

Key: Hb^A = unaffected allele; Hb^S = affected allele

Worked example: Sickle-cell anemia

11. Predict the phenotype of the offspring produced for a cross between a female who has sickle-cell anemia and a man with unaffected alleles.

Solution

Knowing that sickle-cell anemia is an autosomal codominance disease, the following key is to be written:

* Unaffected allele: Hb^A
* Affected allele: Hb^S

The mother has sickle-cell anemia, so her genotype is $Hb^S Hb^S$

The father is unaffected, so his genotype is $Hb^A Hb^A$

Parent generation (F_0)

Phenotype: affected mother × unaffected father

Genotype: Hb^SHb^S × Hb^AHb^A

Gametes: Hb^S, Hb^S Hb^A, Hb^A

First generation (F₁)

	Hb^A	Hb^A
Hb^S	Hb^S Hb^A	Hb^S Hb^A
Hb^S	Hb^S Hb^A	Hb^S Hb^A

F₁ Genotype: all Hb^{S}Hb^A

Phenotype: Half of the blood is unaffected

100% of children will have half of their blood unaffected and the other half sickled.

Question

13 Two parents did not have sickle-cell anemia but one of their children was completely affected with sickle-cell anemia. Predict the genotype of both parents.

DP ready Theory of knowledge

Causality and correlation

A link has been observed between sickle-cell anemia and the prevalence of malaria. Is there a way to tell if this is a causal link or simply a correlation? Studies have shown that there is a correlation between the infection with malaria and having sickle-cell anemia. Can we assume that having sickle-cell anemia increases the chance of becoming infected with malaria? Could it be a cause of malaria?

DP ready Approaches to learning

Simple monohybrid cross problems fall into four categories:

1. Autosomal (possible genotypes: AA, Aa, aa)

2. Codominance (possible genotypes: C^{B}C^B, C^{W}C^W, C^{B}C^W)
   • To be recognized when the characteristics of the offspring are a mixture of the characteristics of the parents

3. Multiple alleles (see blood type example)

4. Sex linkage (Possible genotypes: X^{N}X^N, X^{N}X^n, X^{n}X^n, X^{N}Y, X^{n}Y)

Hemophilia

Hemophilia is an example of a sex-linked recessive disease. People with this disease do not have the proteins that are needed for blood clotting. As a result, they will bleed for a longer period of time when injured. In severe cases, bleeding in joints, muscles, brain or other organs might take place resulting with serious implications. This disease is caused by a gene that is carried on the X chromosome (table 10).

Table 10. Possible genotypes and phenotypes for hemophilia

Genotype	Phenotype
$X^H Y$	Unaffected male
$X^h Y$	Affected male
$X^H X^H$	Unaffected female
$X^H X^h$	Unaffected female (carrier)
$X^h X^h$	Affected female

Key: X^H = unaffected allele; X^h = affected allele

Worked example: Hemophilia **WE**

12. What phenotypes would you predict from a cross between a homozygous unaffected female with an affected male for hemophilia.

Solution

Knowing that hemophilia is a recessive sex-linked disease, the following key is to be written:

- Dominant allele: X^H = unaffected allele
- Recessive allele: X^h = affected allele

A homozygous unaffected female will have a genotype $X^H X^H$

An affected male will have a genotype $X^h Y$

Parent generation (F_0)

Phenotype: unaffected female × affected male

Genotype: $X^H X^H \times X^h Y$

Gametes: X^H, X^H X^h, Y

First generation (F_1)

	X^h	Y
X^H	$X^H X^h$	$X^H Y$
X^H	$X^H X^h$	$X^H Y$

F_1 Genotype: $X^H X^h$, $X^H Y$

Phenotype: all females are unaffected but carriers; all males are unaffected.

Pedigree charts

A *pedigree chart* is a diagram that shows the genetic history of a family over several generations (figure 22).

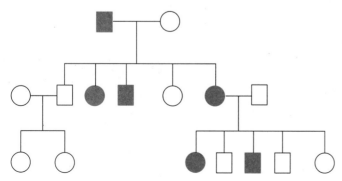

Figure 22. Pedigree chart

Key term

Pedigree charts show the genetic history of a family over several generations. They can be used to uncover the pattern of inheritance of genetic disorders.

In a pedigree chart note that:

- Squares represent males
- Circles represent females
- Shaded symbols represent affected individuals
- Unshaded symbols represent unaffected individuals.

A pedigree chart is used to deduce the pattern of inheritance of genetic diseases. They can show us if the disorder is sex-linked or not.

- The disorder is sex-linked if the number of affected males in the pedigree is much more than affected females.
- The disorder is autosomal if the number of affected males in the pedigree is similar to the number of affected females.

Pedigree charts can also show us if the disorder is the result of dominant or recessive genes.

- The disorder is dominant if one of the parents of an affected child is also affected.
- The disorder is recessive if the both parents are unaffected, but a child is affected.

If we look at figure 22, we can deduce the following:

- This is an autosomal disorder because the ratio between affected males and females is equal.
- This is a dominant disorder because in all the groups of parents with affected children, one parent is also affected.

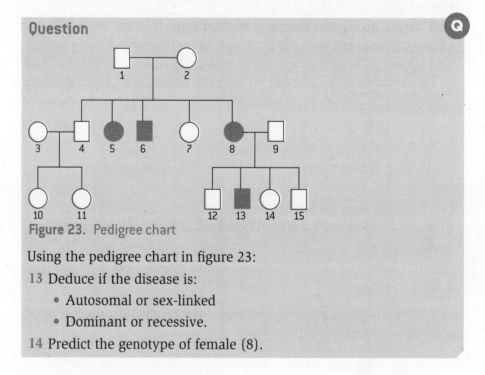

Question

Figure 23. Pedigree chart

Using the pedigree chart in figure 23:

13 Deduce if the disease is:
  - Autosomal or sex-linked
  - Dominant or recessive.

14 Predict the genotype of female (8).

5.4 Biotechnology

Simply put, biotechnology is technology that makes use of biological organisms, processes or systems. Important examples include *DNA profiling*, *gene transfer* and genetic modification.

DNA profiling

DNA profiling is a technique where a sample of DNA is copied and analysed to identify individuals by characteristics of their DNA.

Key term

The technique of **DNA profiling** takes a DNA sample and copies it so that it can be analysed to identify individuals by characteristics of their DNA.

Each individual has unique DNA that is different from anyone else's. DNA profiling is used in medical and genetic screening, as well as part of criminal investigations.

Stages of DNA profiling

There are a number of stages involved in the production of a DNA profile (figure 24).

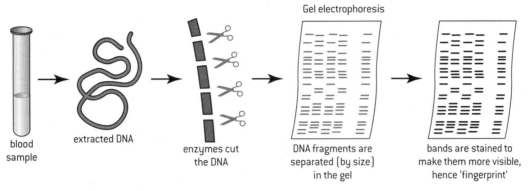

Figure 24. Stages of DNA profiling

1. A DNA sample is collected. A suitable sample can be blood, semen or any biological tissue.
2. The DNA is copied using polymerase chain reaction (PCR), which is a technique that amplifies the amount of DNA using DNA polymerase enzyme.
3. The large amount of DNA produced is cut up into small fragments using restriction enzymes.
4. Gel electrophoresis is then used to separate fragments of DNA according to their size. This gives a pattern of bands on a gel which is unlikely to be the same for two individuals.
5. The DNA banding patterns are compared with other samples of DNA.

Uses of DNA profiling

- Paternity testing: DNA profiling can be carried out, using a mouth swab, to determine whether an individual is the biological parent of a child. One looks for similarities in the DNA profile between the child and the tested individual.
- Forensic investigations: Often tiny samples of DNA can be found such as a drop of blood or saliva to produce a DNA profile that could be compared with profiles of potential suspects (figure 25). If both bands are identical, then DNA belongs to the suspect.

DP link

Polymerase chain reaction (PCR) will be studied in **3.5 Genetic modification and biotechnology** in the IB Biology Diploma Programme.

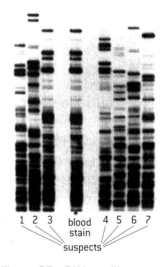

Figure 25. DNA profile

Question

15 Research and describe other uses of DNA profiling.

DP ready **Theory of knowledge**

In the light of discovering new evidence in science, particularly with regards to DNA, how much can we depend on the reliability of using DNA for securing convictions in legal cases?

Gene transfer

Genetic modification is carried out by *gene transfer* between species. When genes are transferred between species, the gene codes for the formation of the same protein. This is because the genetic code is **universal** and will code for the same amino acids in all species.

Examples of genetically modified crops

Salt-resistant tomatoes are genetically modified to tolerate high levels of salt in the soil, which allows them to grow in regions with high salinity.

Maize crops are genetically modified to synthesize Bt toxins which kill insects. This is done because maize crops are often destroyed by insects. The gene responsible for the production of Bt toxin is transferred from bacteria to maize crops.

There are some benefits associated with genetic modification of crops such as increasing crop yield and using less land for production. However, some potential risks might be associated with genetic modification such as the spread of such genes into other plants by cross-pollination. It also could be harmful to humans.

> **DP ready   Nature of science**
>
> Studies have shown that there are many risks associated with genetically modified plants and animals. For instance, risks to monarch butterflies from Bt maize crops were assessed, and data have shown that there is a correlation between the drop in the number of monarch butterflies and the use of Bt maize crops. Scientists are assessing such risks and, as a result of the assessment, serious consequences have been adopted in some areas. For example, some countries have banned the use of genetically modified organisms until they are proven to be safe. The risks associated with any scientific research must be assessed fully to determine whether or not to proceed with the research.

Figure 26. Maize may be genetically modified to improve resistance to insects

Use of gene transfer to produce insulin

It is possible to insert the human gene which codes for insulin into a host cell such as *E. coli*, and allow it to clone and produce insulin to be used for diabetes (figure 27). This can be done as follows:

1. The gene that codes for insulin is extracted from a human pancreatic cell.

2. Restriction enzymes are used to cut the gene and produce single-stranded sections known as "sticky ends".

3. The same restriction enzymes are used to cut a plasmid that is removed from a bacterium cell and produce matching sticky ends with the gene.

4. The matching sticky ends of the gene and the plasmid are joined using the enzyme DNA ligase to form what is known as a *recombinant plasmid*.

5. The recombinant plasmid is then inserted into a host cell such as *E. coli*, which is allowed to clone and produce insulin.

6. The desired gene (in the host cells) produces insulin which can be collected for use by diabetics.

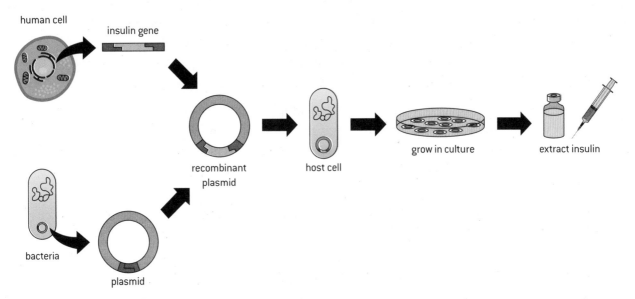

Figure 27. Gene transfer

Question

16 List what is required for gene transfer.

Chapter summary

In this chapter, you have learned about the material of inheritance carried in chromosomes, the process of meiosis, and inheritance through monohybrid crosses including autosomal, sex linkage, codominance and multiple alleles. Make sure that you have a working knowledge of the following concepts and definitions:

- ☐ Chromosomes are made up of many genes that control all our characteristics. They occur in pairs of homologous chromosomes in all body cells, except in sex cells.
- ☐ A gene is a small section of DNA that codes for a specific protein in your body. DNA carries the genetic code that determines the characteristics of a living organism.
- ☐ DNA is made of coding and non-coding regions. The coding region codes for the formation of a protein while the non-coding regions are not involved in protein synthesis.
- ☐ A karyogram is a photograph that shows the chromosomes of an organism in homologous pairs of decreasing length. Karyograms can be used to deduce sex and determine any chromosomal abnormalities.
- ☐ A mutation is any change to the base sequence of a gene in an organism.
- ☐ Meiosis occurs in the sex organs (testes and ovaries). It results in the formation of four haploid cells from a diploid cell.
- ☐ Non-disjunction occurs when the chromosomes do not separate properly during meiosis. This results in the production of gametes that either have missing or extra chromosomes.
- ☐ The inheritance of traits is controlled by genes. A gene has different forms called alleles.
- ☐ A dominant allele is an allele that affects the phenotype when present in the homozygous or heterozygous state. Dominant alleles mask the effects of recessive alleles.
- ☐ A recessive allele is an allele that affects the phenotype only when present in the homozygous state.
- ☐ A test cross is testing an unknown by crossing it with a homozygous recessive.
- ☐ Inheritance of ABO blood groups in humans is an example of codominance and multiple alleles.
- ☐ The two sex chromosomes determine gender, and they are the X and the Y chromosomes. Females have two X chromosomes whereas males have one X and one Y chromosome.
- ☐ Sex linkage is a trait that is controlled by a gene located on the sex chromosome and is associated with gender.
- ☐ DNA profiling is a technique where a sample of DNA is copied and analysed to identify individuals by characteristics of their DNA. Each individual has a unique DNA that is different from anyone else's.
- ☐ Genetic modification is carried out by gene transfer between species.

Additional questions

1. Identify factors that affect the acceptance of new scientific ideas.
2. Explain why sex linkage disorders are more common in males than females.
3. If the haploid number of a species is 12, determine how many chromatids will there be in metaphase I in a dividing diploid cell.
4. Explain what happens in crossing over.
5. A woman who is a carrier for colour blindness and a man who has unaffected vision have a child. Determine the probability that the child will have colour blindness.
6. Predict the genotypes in the offspring of a homozygous male with blood group A and a female with blood group B.

7. In dogs, wire hair is due to a dominant gene (W) and smooth hair is due to its recessive allele (w). If a heterozygous wire-haired dog is mated with a smooth-haired dog, what type of offspring could be produced?

8. From the pedigree chart in figure 28, deduce if the condition is caused by a dominant or a recessive allele, and if it is sex-linked or autosomal.

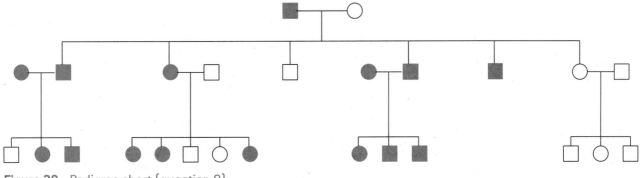

Figure 28. Pedigree chart (question 8)

9. Explain what a recombinant plasmid is.

" Keep steadily in mind that each organic being is striving to increase at a geometrical ratio; that each at some period of its life, during some season of the year, during each generation or at intervals, has to struggle for life, and to suffer great destruction. When we reflect on this struggle, we may console ourselves with the full belief, that the war of nature is not incessant, that no fear is felt, that death is generally prompt, and that the vigorous, the healthy, and the happy survive and multiply. "

Charles Darwin, 1859

Chapter context

Ecology is the study of relationships between living organisms, and between organisms and their environment. An **ecosystem** is a **community** of living organisms interacting with the non-living things in their environment. Nutrients are recycled in an ecosystem, unlike energy which must be continuously supplied due to its loss. The **greenhouse effect** is due to the greenhouse gases present in the atmosphere. The theory of **evolution** and **natural selection** discusses how species change over geographical time. **Classification** is the process by which living organisms are sorted out into different groups based on their similarities.

Learning objectives

In this chapter you will learn about:

→ **ecosystems** and what makes up an ecosystem from **biotic** and **abiotic** **factors**

→ the types of living organisms based on the mode of nutrition (**autotrophs** and **heterotrophs**)

→ the transfer of energy and nutrients in ecosystems through **food chains**

→ the causes and consequences of the **greenhouse effect**

→ the theory of **evolution** and **natural selection**

→ **classification** of living organisms.

🔑 Key terms introduced

→ Ecosystems, community, biotic factors and abiotic factors

→ Species, habitats and population

→ Autotrophs (producers) and heterotrophs

→ Consumers, parasites, detritivores and saprotrophs

→ Internal and external digestion

→ Food chain and trophic levels

→ Food web

→ Pyramid of energy

→ Greenhouse gases and the greenhouse effect

→ Evolution and natural selection

→ Speciation and gradual divergence

→ Dichotomous key

🔑 Key term

Ecosystem is the term used to describe the different living organisms living together and interacting with each other and with the abiotic environment.

Community describes the populations of different species living and interacting with each other in a defined environment.

Biotic factors are the living organisms in an ecosystem.

Abiotic factors are the non-living organisms such as soil, water and sunlight in an ecosystem.

6.1 Ecosystems

An *ecosystem* is a *community* of living organisms interacting with the non-living things in their environment. It could be as small as a tree or as large as a desert. An ecosystem could be natural such as a lake, or artificial such as an aquarium.

Components of an ecosystem

An ecosystem consists of:

- *Abiotic factors,* which are the non-living things such as soil, atmosphere, heat, sunlight and water
- *Biotic factors,* which are the different living organisms interacting with those abiotic factors.

An ecosystem includes different *species*. Species are groups of individuals that have the potential to interbreed and produce fertile offspring. If members of two closely related species interbreed and produce offspring, the hybrids will likely be sterile (infertile).

Mules are the offspring of a female horse and a male donkey, which are two different species with a different number of chromosomes.

The area in which species live is called a *habitat*. The different species living together in the same area at the same time are called *populations*. The different populations make up a community.

Worked example: Ecosystems

1. From figure 1, identify the following: an example of a species, population, community and ecosystem.

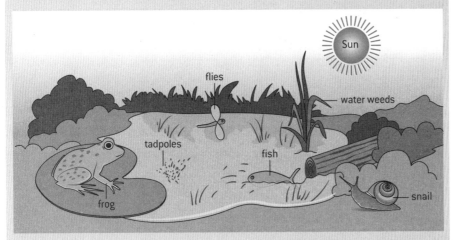

Figure 1. A pond ecosystem

Solution

Species: frogs.

Population: the same species such as the group of frogs at the pond.

Community: the different populations of fish, snails and frogs all interacting together.

Ecosystem: the community of all organisms interacting with the Sun, soil and water.

Measuring population size

Quadrat sampling is one method that can be used to measure the size of different populations within an ecosystem. The results obtained from the sampled area can then be used to estimate the size of the population.

Practical skills: Quadrat sampling

Quadrats are used to estimate the size of a population in an ecosystem. A quadrat is a square made of a wire that is usually 1 m² and may include smaller squares inside (figure 2). Some may include 25 squares (5 × 5) or 100 squares (10 × 10). Quadrats are usually used for sessile (non-moving) organisms such as plants or slow-moving organisms such as snails.

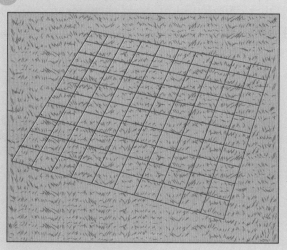

Figure 2. A quadrat

How to use quadrats

a) Mark out the area to be sampled.

b) Place the quadrat randomly on the ground within the area identified.

c) Count the number of individual organisms (for example snails or dandelions) inside each quadrat.

d) The data collected can be used to estimate the population size within the habitat as follows:

- Calculate the average number of organisms per quadrat.
- Calculate the whole field area.
- Calculate the estimated population size of dandelions in the field.
- The greater the number of different quadrat samples used, the more reliable and valid the results are.

Worked example: Calculations involving quadrats

2. Calculate the estimated population size of snails in a field of an area 100 m², using a 1 m² quadrat, knowing that the number of snails counted for 10 quadrats was as follows:

3, 2, 4, 1, 0, 0, 2, 1, 2, 0

Solution

Calculate the average number of snails per quadrat:

$$\frac{\text{Sum of the number of snails}}{\text{Number of quadrats}} = \frac{15}{10} = 1.5 \text{ per } 1 \text{ m}^2 \text{ quadrat}$$

Estimated population size of snails in 100 m² = 1.5 × 100

= 150 snails per 100 m²

Practical skills: The chi-squared test

You may use the chi-squared (χ^2) test with data obtained by quadrat sampling to test if there is a significant association between two organisms.

If two organisms are present in the same habitat, they may show a negative or positive association.

A negative association means that the two organisms do not interact with each other and their presence does not depend on each other. This shows that the two species are not associated and therefore, their distribution is random.

A positive association means that the two organisms interact with each other and their presence depends on each other. This could be through a symbiotic relationship or a predator–prey relationship. This means that the two species are associated and therefore, their distribution is not random.

To start the chi-squared test, you need to obtain data by quadrat sampling where the following are identified:

- The number of quadrats where only organism A is present.
- The number of quadrats where only organism B is present.
- The number of quadrats where both organisms are present.
- The number of quadrats where both organisms are absent.

A chi-squared test may be conducted by following the steps below:

a) Identify **hypotheses** (null versus alternative)

b) Construct a **contingency table of observed frequencies**

c) Construct a **contingency table of expected frequencies**

d) Apply the **chi-squared formula** (χ^2)

e) Determine **the degree of freedom** (df)

f) Identify the P value and compare it with the critical value (should be $P < 0.05$).

These steps are carried out in the following worked example.

Worked example: The chi-squared test

3. A student carried out quadrat sampling and obtained the following results (table 1).

Table 1. Student's quadrat sampling data

Number of quadrats	Organisms present
5	Both organisms
14	Gaper clam
19	Butter clam
8	None

Conduct the chi-squared test to find out if there is a significant association between the two types of clams.

Solution

a) Identify hypotheses:

- **Null hypothesis (H_0):** There is **no** association between the butter clam and the gaper clam.
- **Alternative hypothesis (H_1):** There **is** an association between the butter clam and the gaper clam.

b) Construct a contingency table of **observed frequencies** (table 2). Compare this table to table 1; can you see which values in table 1 correspond to the values in table 2?

Table 2. Contingency table of observed frequencies for butter and gaper clams

		Butter clam		
		Present	Absent	Total
Gaper clam	Present	5	14	19
	Absent	19	8	27
	Total	24	22	46*

*46 is the grand total.

c) Construct a contingency table of **expected frequencies** (table 3):

Use the following formula to calculate the expected frequency for each:

$$\text{Expected frequency} = \frac{(\text{row total} \times \text{column total})}{\text{grand total}}$$

Table 3. Contingency table of expected frequencies for butter and gaper clams

		Butter clam		
		Present	Absent	Total
Gaper clam	Present	$\dfrac{(19 \times 24)}{46} = 9.9$	$\dfrac{(19 \times 22)}{46} = 9.1$	19
	Absent	$\dfrac{(27 \times 24)}{46} = 14.1$	$\dfrac{(27 \times 22)}{46} = 12.9$	27
	Total	24	22	46

d) Apply the chi-squared formula (χ^2):

Use the formula below to calculate a statistical value for the chi-squared test:

$$\chi^2 = \sum \frac{(O - E)^2}{E}$$

Where Σ = sum; O = observed frequency; E = expected frequency.

These calculations can be broken down for each part of the distribution pattern to make the final summation easier (table 4).

Table 4. Calculations for the chi-squared (χ^2) test

		Butter clam		
		Present	Absent	Total
Gaper clam	Present	$\dfrac{(5 - 9.9)^2}{9.9} = 2.43$	$\dfrac{(14 - 9.1)^2}{9.1} = 2.64$	19
	Absent	$\dfrac{(19 - 14.1)^2}{14.1} = 1.70$	$\dfrac{(8 - 12.9)^2}{12.9} = 1.86$	27
	Total	24	22	46

The sum of the individual values is then calculated:

$\chi^2 = (2.43 + 2.64 + 1.70 + 1.86) = 8.63$.

e) Determine the degree of freedom (df):

You can calculate the degree of freedom from the table of frequencies as follows:

df = (number of rows − 1) × (number of columns − 1)

df = (2 − 1) × (2 − 1) = 1

f) Identify the *P* value and compare it with the critical value:

Use the chi-squared distribution table (table 5) to determine if the value obtained (8.63) is significant or not. Look for the df that is equal to 1, and determine the *P* value of χ^2 (8.63). The closest value to 8.63 is the value of 6.635. Therefore, the *P* value of 8.63 is less than 0.01 (*P* < 0.01).

Table 5. Chi-squared distribution table

df	*P* values for Chi-square (χ^2) distribution							
	0.90	0.75	0.50	0.25	0.10	0.05	0.025	0.01
1	0.016	0.102	0.455	1.320	2.706	3.841	5.024	6.635

→ *statistically significant*

Any result that is more than the critical value (3.841), where the *P* value is less than 0.05, is considered significant. Since the *P* value of 8.63 is less than 0.01, it is considered significant. This means that the null hypothesis is rejected, and the alternative hypothesis is accepted. Therefore, we can say that there is an association between the butter clam and the gaper clam.

Question

1 What are the two components of an ecosystem?

2 Explain why quadrat sampling is not an effective method in estimating the population size of all living organisms.

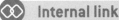 **Internal link**

P values were discussed in **4.2 Plant transport** in this book.

6.2 Energy and nutrient transfer: modes of nutrition

For an ecosystem to survive, the biotic and abiotic factors need energy and nutrients. Energy in the form of sunlight is continuously available, though the supply varies. For example, there is less sunlight available in winters than in summers because there are fewer hours of daylight. In contrast, the supply of inorganic nutrients in an ecosystem is finite, meaning that the supply is being constantly cycled through the ecosystem.

There are a number of ways in which the organisms in an ecosystem obtain the necessary energy and nutrients they need. These methods are known as modes of nutrition. Living organisms can be divided into two main categories, autotrophs and heterotrophs, based on the way they obtain their food.

Autotrophs

The term *autotroph* is derived from two Greek words: **auto** which means "self" and **trophe** which means "nutrition". Autotrophs are organisms that synthesize their organic molecules from simple inorganic nutrients. The inorganic nutrients used by autotrophs are from the abiotic environment. Most plants and algae are autotrophic, which means that they synthesize their own food through photosynthesis. Autotrophic organisms are also known as *producers*.

 DP link

The chi-squared test will be studied in detail in **4.1 Species, communities and ecosystems** in the IB Biology Diploma Programme.

 Internal link

Energy flow through the ecosystem will be discussed in **6.3 Energy transfer in ecosystems**.

The recycling of inorganic nutrients and the carbon cycle will be discussed in **6.4 Nutrient transfer in ecosystems**.

Looking for patterns, trends and discrepancies

Although most plants and algae are autotrophic, some are not. A small number of plants and algae are not considered autotrophs and do not fit into this trend. This is because they lack chloroplasts and therefore cannot carry out photosynthesis. They usually live on other plants and so are considered parasitic (for example, plants in the Rafflesia or Corpse Flower family are parasites). Do parasitic plants and algae falsify the theory that plants and algae are autotrophs? Since the number of parasitic plants and algae is relatively small compared to the autotrophic ones, ecologists consider plants and algae as autotrophs, with some exceptional species as parasitic.

Key term

Autotrophs (producers) are organisms that synthesize their organic molecules from simple inorganic nutrients.

Heterotrophs are organisms that obtain organic nutrients from other organisms.

Key term

Consumers are heterotrophs that obtain their food from other living organisms; they can be herbivores, carnivores or omnivores.

Parasites are heterotrophs that obtain their food from a host organism.

Detritivores are heterotrophs that obtain their food from detritus by internal digestion.

Saprotrophs are heterotrophs obtain their food from dead organisms by external digestion.

Detritus is the non-living organic matter such as animal remains, waste products and other organic matter that falls onto the soil from the surroundings.

Internal digestion occurs when the organism secretes digestive enzymes to break down food inside the body.

External digestion occurs when the organism secretes digestive enzymes to break down the food outside the body and then absorbs the products of digestion.

Heterotrophs

The term *heterotroph* is derived from two Greek words: **hetero** which means "other" and **trophe** which means "nutrition". Heterotrophs are organisms that obtain organic nutrients from other organisms. Depending on the method of food intake, heterotrophs may be *consumers, parasites, detritivores* and *saprotrophs*.

Consumers are heterotrophs that obtain their organic nutrients by ingesting living organisms such as plants or animals. Consumers could be herbivores, carnivores or omnivores.

• Herbivores, for example zebras, feed on producers (such as grass).

• Carnivores, for example lions, feed on other consumers (such as zebras).

• Omnivores, for example chimpanzees, feed on a combination of both producers and consumers.

Parasites are heterotrophs that live on or inside other living organisms (called hosts) and obtain their food from them. Tapeworms are an example of a parasite.

Detritivores (for example, an earthworm) are heterotrophs that obtain organic nutrients from *detritus* (non-living organic matter) by *internal digestion*.

Saprotrophs are heterotrophs that obtain organic nutrients from dead organisms by *external digestion*. Saprotrophs are also known as decomposers because they have a role in breaking down organic material. Bacteria and fungi are examples of saprotrophs.

Some organisms use a combination of different modes of nutrition and can act as both autotrophs and heterotrophs depending on the surrounding conditions. These organisms are known as *mixotrophs*.

For example, euglena is an algae that can act as both an autotroph and a heterotroph (figure 3). In sufficient light, it will act as an autotroph and can photosynthesize to make its own food. In low light, it will act as a heterotroph and ingest food particles by phagocytosis.

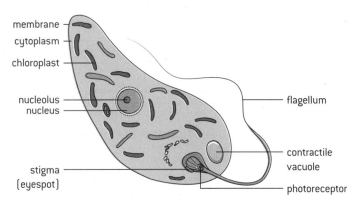

Figure 3. Euglena

Another example is the Venus flytrap (figure 4), a plant that can act as an autotroph like most plants and can photosynthesize to make its own food. However, due to the nutrient poor soil of the wetlands where it lives, it traps and digests both insects and spiders just like an autotroph.

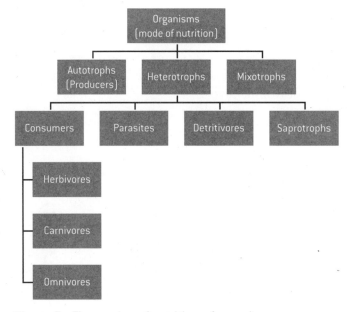

Figure 4. Venus fly trap

Figure 5 summarizes the various modes of nutrition discussed in this section.

Figure 5. The modes of nutrition of organisms

Question

3   Classify the following species into autotrophs, consumers, detritivores or saprotrophs:

   a)   Millipedes that eat decaying plant and animal matter in the soil.

   b)   A deer that eats plants.

   c)   Phytoplankton present in the sea that use the sunlight to make their own food.

6.3 Energy transfer in ecosystems

Food chains, *food webs* and *pyramids of energy* are ways of describing how and where energy moves through an ecosystem.

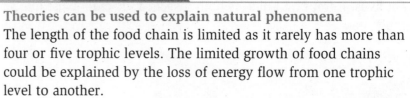

DP ready   **Nature of science**

Theories can be used to explain natural phenomena
The length of the food chain is limited as it rarely has more than four or five trophic levels. The limited growth of food chains could be explained by the loss of energy flow from one trophic level to another.

Key term

A **food chain** shows the movement of energy from one organism to the next.

The **trophic level** is the position of an organism in the food chain.

Food chains

Food chains show the movement of energy from one organism to the next. The direction of the energy flow is represented by an arrow. The position of the organism in the food chain is called the *trophic level*. The term "trophic" comes from a Greek origin which means "to feed". The trophic level of an organism depends on the type of food it feeds on:

- Producers make their own food and they are the beginning of any food chain.

- Primary consumers are herbivores that feed on producers.

- Secondary consumers are carnivores or omnivores that feed on herbivores.

- Tertiary consumers are carnivores or omnivores that feed on secondary consumers.

Producers, primary consumers, secondary consumers and tertiary consumers are examples of trophic levels.

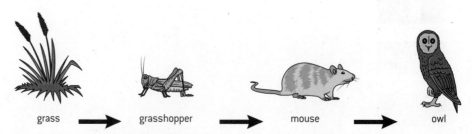

grass → grasshopper → mouse → owl

Figure 6. An example of a food chain

Look at figure 6. In this food chain, grass is the producer. It gets energy from the Sun and uses it to make its own food. The grasshopper is a primary consumer as it eats grass. The mouse is a secondary consumer because it eats the grasshopper. The owl is a tertiary consumer because it feeds on the mouse.

Food webs

In most communities, organisms eat more than one organism which results in several interacting food chains. A food web is a diagram that shows the relationships between organisms in a community. The arrows in the food web show the direction of the energy flow (figure 7).

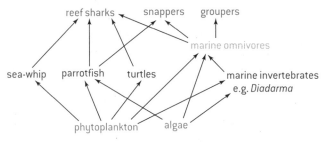

Figure 7. An example of a food web

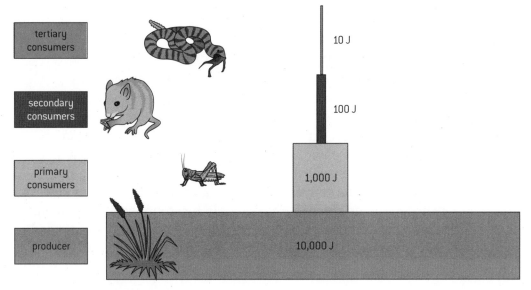

> **Key term**
>
> A **food web** is a diagram that shows the relationships between the organisms in a community and the direction of the energy flowing through it.

Question

4   In the food web in figure 7, identify:

   a) A producer

   b) A primary consumer

   c) A secondary consumer

   d) A tertiary consumer

   e) An organism that can be both a secondary and a tertiary consumer.

Pyramid of energy

The amount of energy flow in a food chain is represented by a pyramid of energy. Each bar in the pyramid represents the amount of energy at each trophic level in the food chain. Producers form the base of any energy pyramid (figure 8).

> **Key term**
>
> A **pyramid of energy** shows the amount of energy in a food chain.

Figure 8. An example of a pyramid of energy. In this example, each bar represents the energy per unit area per year of a trophic level. The unit is joule per metre squared per year ($J\ m^{-2}\ yr^{-1}$).

Energy flow in a food chain

Sunlight is the source of energy on Earth. Energy enters a food chain from sunlight through producers which convert the light energy into chemical energy in organic compounds by photosynthesis. Chemical energy in carbon compounds flows through trophic levels by means of feeding. Energy transformations are never 100% efficient.

In the pyramid of energy in figure 8, notice the proportions of each bar drawn. Each bar is smaller than the one below. This is because only 10% of energy is transferred from one trophic level to the next and the remaining 90% is lost. Energy loss between trophic levels is mainly due to heat loss by cellular respiration. Therefore, ecosystems require a continuous supply of energy to replace energy lost as heat. Energy is passed to saprotrophs and detritivores in dead organic matter.

Worked example: Calculations involving energy flow

4. Calculate the amount of energy transferred to the secondary consumer knowing that the energy of producers is $20\,000$ J m^{-2} yr^{-1}.

Solution

Energy transferred to each trophic level = 10%.

Energy transferred from producers to primary consumer = 10% of

original $= \dfrac{10}{100} \times 20\,000$

$= 2,000$ J m^{-2} yr^{-1}

Energy transferred from primary consumer to secondary consumer

$= \dfrac{10}{100} \times 2000 = 200$ J m^{-2} yr^{-1}

Question

5 In the following food chain, calculate the amount of energy (in kJ) transferred to tuna fish knowing that the energy in phytoplankton is 150 000 J m^{-2} yr^{-1}.

- Phytoplankton → Shrimp → Tuna fish → Sharks

6.4 Nutrient transfer in ecosystems

Nutrients consisting of elements such as carbon, nitrogen and phosphorus are required by an organism for growth and metabolism. Since there is a limited supply of inorganic nutrients, they must be recycled among organisms. Saprotrophs recycle nutrients and return them to the environment to be reused. Autotrophs obtain inorganic nutrients available in the ecosystem such as water and carbon dioxide and convert them to organic nutrients such as glucose. Organic nutrients are then transferred to heterotrophs through food. Saprotrophic bacteria and fungi decompose dead organic matter and release nutrients back to the soil as inorganic nutrients. This ensures the recycling of nutrients within an ecosystem.

Sustainability of ecosystems

Ecosystems tend to stay sustainable for a long period of time by ensuring that there is a continuous supply of energy; this energy is provided by sunlight. Recycling of nutrients and waste products by saprotrophs is also needed for an ecosystem's sustainability.

Practical skills: Prepare a mesocosm to model a sustainable ecosystem

A mesocosm is a model of a larger ecosystem. The term mesocosm comes from the term "meso" which means "medium" and "cosm" which means "world".

When setting up a mesocosm, you can either have it as a closed or open ecosystem. Closed ecosystems sealed in glass are preferred as they prevent the entry and exit of external matter but at the same time allow light to enter and leave.

Mesocosms allow us to examine factors that may affect natural ecosystems under controlled conditions. Therefore a mesocosm is considered a valuable tool to study biotic and abiotic features in ecological research. For example, you can set up a mesocosm to examine the effect of increased levels of carbon dioxide on specific types of organisms. Manipulation of environmental factors in a mesocosm enables us to observe the effect of such factors and therefore build an understanding of the correlations between biotic and abiotic factors. However, the limited space in a mesocosm does not allow organisms to behave in the same way that they behave in a natural environment, which may affect results obtained.

You can create a simple mesocosm by using a glass jar to which you add a mixture of biotic and abiotic factors, and ensure it is tightly sealed and placed close to sunlight (figure 9). The biotic and abiotic factors may include:

- Biotic factors—these include different living organisms from plants (example, cactus) and animals (example, worms, etc.)
- Abiotic factors—these may include non-living organic matter such as dead leaves, humus, soil, pebbles and water.

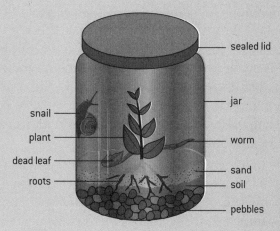

Figure 9. An example of a mesocosm

 DP link

Ecosystem sustainability will be studied further in **4.1 Species, communities and ecosystems** in the IB Biology Diploma Programme.

The carbon cycle

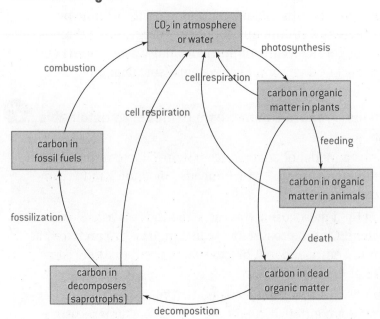

Figure 10. The carbon cycle

Carbon is continually available in ecosystems because of carbon cycling (figure 10). Carbon is present in the atmosphere or in water in the form of carbon dioxide. Sources of carbon dioxide include:

- **Cellular respiration**: Carbon dioxide is produced by the cellular respiration of all living organisms. Carbon dioxide diffuses out of organisms into water or the atmosphere.

- **Methane oxidation**: Methane is a gas that is produced as a waste product by *methanogenic archaea*, which are organisms that live in anaerobic conditions such as wetlands and the digestive tract of some living organisms such as cows and humans. When methane is produced, some of it is oxidized in the atmosphere into carbon dioxide and water.

- **Combustion (burning) of fossil fuels (fossilized organic matter)**: When saprotrophs decompose dead organic matter, decomposition is not fully completed due to various conditions such as anaerobic or acidic conditions. The building up of large amount of partially decomposed dead organic matter forms peat. When peat from past geological eras is exposed to high pressure and heat over millions of years, it is converted into fossil fuels including coal, oil or gas. The combustion (burning) of such fossil fuels releases carbon dioxide back into the atmosphere. The hard parts of reef-building corals are composed of calcium carbonate and may become fossilized in limestone rock.

Carbon dioxide in the atmosphere and in water is then transferred to producers by photosynthesis. Carbon is transferred from one living organism to another through feeding. When plants and animals die, decomposers (saprotrophs) decompose the organic dead matter to release carbon back into the soil. Geological processes have converted dead organic matter from past geological eras into fossil fuels.

Carbon dioxide, and the way it is continually recycled by the carbon cycle, plays a major role in the *greenhouse effect* and climate change.

Key term

Methanogenic archaea are organisms that live in anaerobic conditions such as wetlands and the digestive tract of some living organisms such as cows and humans.

DP link

Carbon cycling in water and the atmosphere will be studied in detail in **4.3 Carbon cycling** in the IB Biology Diploma Programme.

6.5 Climate change

The concentration of the atmospheric "greenhouse" gases affects the Earth's climates. Although the greenhouse effect is a natural phenomenon, it is being enhanced by human activities.

The greenhouse effect

The *greenhouse gases,* which include carbon dioxide, water vapour, methane and nitrogen oxides, in combination with radiation from the Sun, cause the *greenhouse effect* (figure 11). This natural phenomenon occurs when the Sun's radiation is trapped by the Earth's atmosphere increasing the Earth's temperature until it reaches a level that is suitable for life on Earth. Carbon dioxide and water vapour are the most significant greenhouse gases because of their high concentration in the atmosphere. The effect of any greenhouse gas is determined by its ability to absorb longwave radiation and its concentration in the atmosphere.

Key term

The **greenhouse gases**, which include carbon dioxide, water vapour, methane and nitrogen oxides, are so-called because they cause the greenhouse effect.

The **greenhouse effect** is a natural phenomenon caused by the greenhouse gases. The greenhouse effect keeps the Earth's temperature warm and therefore allows life on Earth.

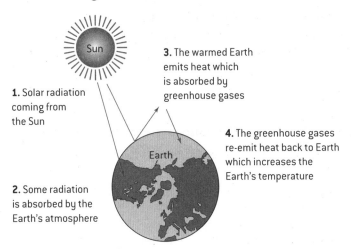

Figure 11. The greenhouse effect

The Sun's radiation along with the greenhouse gases is the cause of the greenhouse effect:

1. The radiation travels through the atmosphere to warm the surface of the Earth.
2. The warmed Earth then emits heat as infrared radiation.
3. The greenhouses gases in the atmosphere absorb the infrared radiation and the atmosphere heats up.
4. The heated atmosphere then re-emits heat back to Earth.

This cycle of heating and reheating is the greenhouse effect, and is necessary to maintain the average mean temperature on Earth. Changes in the levels of greenhouse gases result in changes to the Earth's temperature. A rise in the concentration of greenhouse gases in the atmosphere results in an increase of the greenhouse effect, increasing the Earth's temperature further.

What causes the enhanced greenhouse effect?

The enhanced greenhouse effect is caused by human activities that result in an increased concentration of greenhouse gases in the atmosphere. Such activities include:

- Increased carbon dioxide production due to the burning of fossil fuels and cutting down trees (deforestation)

Internal link

Archaea, including methanogenic archaea, will be studied later in 6.7 Classification.

- Increased methane production due to the leakage of methane from natural gas systems and the raising of livestock. Methane is also produced by anaerobic respiration of methanogenic archaea. Methane is then oxidized to carbon dioxide and water in the atmosphere
- Increased production of nitrogen oxides due to industrial processes.

DP ready | **Nature of science**

When collecting quantitative data, it is important to make sure the data is reliable and accurate. Studies to assess the role of greenhouse gases in climate change rely on accurate measurements of the atmospheric concentration of carbon dioxide and methane. This is essential to evaluate the past and possible impact of future human activities on the concentration of such gases in the atmosphere, which may increase the temperature of the Earth. There are various research stations in different areas of the world to collect data about the concentration of the two gases. Records obtained from Mauna Loa in Hawaii show the longest period of time through which such data was collected.

What are the consequences of the enhanced greenhouse effect?

- Increasing the temperature of the Earth, which causes global warming
- Change in climate
- Extinction of species
- Increase in sea levels due to melting of glaciers.

Question

6   How can we control the increase in the greenhouse effect?

Practical skills: Modelling the greenhouse effect

You can model the greenhouse effect using two flasks, using the following method:

- Fill one flask with air as a control.
- Fill another flask with carbon dioxide. You can collect carbon dioxide gas by generating it from a chemical reaction such as the reaction between sodium bicarbonate and vinegar. Invert a flask filled with water in a pan where this reaction is taking place, and allow the water inside the flask to be displaced by the carbon dioxide produced from the reaction.
- Insert a thermometer (or a data logger) in both flasks and seal both flasks.
- Place both flasks an equal distance from a light bulb and then record the changes in temperature in both flasks over a period of time.
- Compare the results recorded for each flask.

6.6 Evolution

Charles Darwin (figure 12) was a biologist known for his theory of *evolution* and the process of *natural selection*. In his book *On the Origin of Species*, he formulated his theory based on his observations over a five-year voyage around the world where he studied specimens of different species. Darwin's theory of evolution explains how species of living things have changed over geological time—this theory and the process of natural selection later became known as "Darwinism".

The theory of evolution provides an explanation for why species from the past do not exist today and why some of the species we see today did not exist long ago.

The theory of evolution is now accepted by the vast majority of the scientific community, based on strong evidence in its favour. This includes evidence from the *fossil* record, evidence from the selective breeding of animals and evidence from homologous anatomical structures. Despite this evidence, the subject of evolution has been controversial on a social and cultural level.

Figure 12. Charles Darwin (1809–1882)

 Key term

Evolution is the gradual change of features over a long period of time.

 Key term

A **fossil** is any preserved remains or traces of any living thing from a past geological age.

Evidence from fossil records

Fossils show changes over time in many organisms. Based on the fossil records collected, it was noted that many of the fossilized organisms were substantially different from the ones existing in the present day. Some fossilized organisms had similarities with the existing ones but were not identical. This indicated that the species changed over time; some evolved and developed new features while others became extinct.

Evidence from selective breeding of animals

For thousands of years, humans have deliberately bred specific members of species with a desired trait. This usually caused an increase in the frequency of the desired trait and made it more common in successive generations. This shows that selective breeding, which is a type of *artificial selection*, can cause evolution within a population.

Evidence from homologous anatomical structures

Homologous anatomical structures within different species suggest common ancestry. For example, the pentadactyl limb—a limb which normally has five digits—as seen in human hands and feet, bats' wings, cats' paws and whales' fins, suggests that they all had a common ancestor (figure 13).

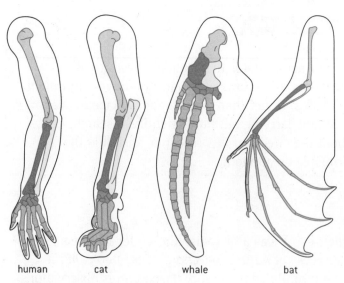

human   cat   whale   bat

Figure 13. Pentadactyl limbs

> **DP ready**   **Nature of science**
>
> Vertebrate limbs are used for different functions such as flying, swimming and walking. Despite their different functions, vertebrate limbs show many common features in the bone structure. This is explained by evolution—the similarity indicates that these vertebrates come from a common ancestor.

Natural selection

Natural selection is the process by which evolution takes place. To understand how natural selection takes place, we need to understand the following key points:

- **Variation**: Individuals in a species show a wide range of genetic variation. Processes which cause variation between individuals in a species include mutation, meiosis and sexual reproduction. This will result in different organisms of the same species showing a range of different physical characteristics.

- **Struggle for survival**: Many species produce more offspring than the limited resources of their environment can support, leading to a struggle for survival.

- **Survival of the fittest**: Due to competition or change in the environment, not all individuals will survive, only the "fittest". Survival of the fittest does not necessarily mean the strongest or the healthiest individuals, but those best adapted to their environment.

- **Adaptation**: Adaptations are characteristics that make an individual suited to its environment and way of life. Some individuals will have characteristics that make them more adapted to the environment than others, and hence these tend to survive and produce more offspring while the less well adapted tend to die or produce fewer offspring. Individuals that reproduce pass on these characteristics to their offspring.

Over time, the favourable characteristics that make a species better adapted to its environment accumulate and a new generation is created.

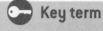

Key term

Natural selection is the process by which evolution takes place due to the genetic variation of individuals in species.

The Galapagos finches

Darwin was fascinated by the animals of the Galapagos islands. He studied 13 different species of finches and he noted how they had different beaks depending on the source of nutrition (figure 14). He suggested that all finches came from a common ancestor. He believed that seed-eating finches reached the island but there was a limited food source for all the finches to survive. Finches which had a different beak shape were better adapted to eat other types of food and therefore, survived and passed on their beak shape to their offspring. Darwin believed that the Galapagos finches evolved into different species by natural selection.

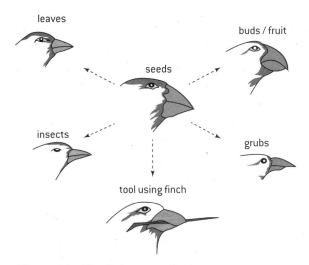

Figure 14. The Galapagos finches

Antibiotic resistance

Antibiotic resistance in bacteria is an example of evolution in response to a change in the environment (figure 15). When antibiotics were discovered and introduced in the medical treatment of humans in the early 20th century, bacteria adapted to this change:

- Bacteria show genetic variation—some may carry a gene that displays resistant properties to a specific antibiotic. This gene is usually acquired by mutation or by a plasmid transferred through conjugation with a bacterial cell that carries the gene.
- The bacteria that have the gene become resistant to the specific antibiotic, while others which do not have the gene die once exposed to the antibiotic.
- A new population of antibiotic-resistant bacteria will be produced.
- Over time a whole new generation of bacteria will become resistant to that specific antibiotic.

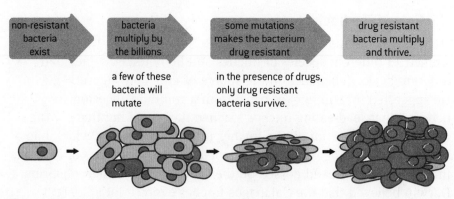

Figure 15. Antibiotic resistance in bacteria

Key term

Speciation is an evolutionary process by which two related populations diverge into separate species and they no longer interbreed.

Gradual divergence occurs when populations of a species slowly transform into separate species by evolution.

Key term

Cladograms are tree diagrams that show how a group of organisms (clades) have evolved from a common ancestor.

A **clade** is a group of organisms that have evolved from a common ancestor.

Speciation

Speciation occurs when the populations become so different that they no longer interbreed. New species arise as a result of:

- Genetic variation, potentially as a result of mutation.
- Isolation of populations; if two populations of a species become geographically separated, then they will likely experience different ecological conditions and hence, *diverge gradually* from one another.

Cladograms are tree diagrams that show how a group of organisms (*clades*) have evolved from a common ancestor (figure 16). Cladograms are used to deduce evolutionary relationships.

Figure 16. An example of a cladogram

Evidence for which species are part of a clade can be obtained from:
- the nucleic acid and base sequences
- the amino acid sequence of a protein.

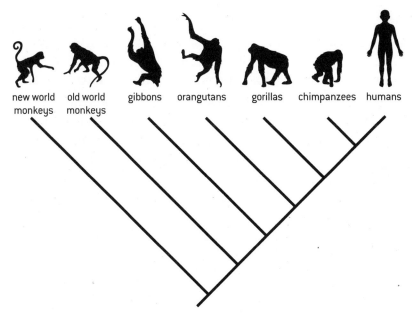

Figure 17. Cladogram of humans and other primates

Cladistics have provided evidence that classification of some species based on their structure did not match with their evolutionary origin. Evidence showed that many species diverged over time and became very different from the original one. This resulted in the reclassification of some organisms. Therefore, cladograms could be used to analyse any evolutionary relationships (figure 17).

Practical skills: How do you analyse a cladogram to deduce evolutionary relationships?

From a cladogram, such as figure 18, you can deduce:

- The common ancestor of descendants
- The time when speciation took place
- If any organisms have a shared history.

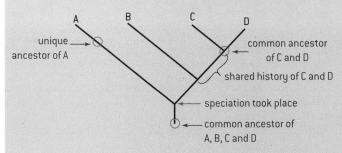

Figure 18. Deducing evolutionary relationships from a cladogram

DP ready   Nature of science

Sometimes the development of new theories falsifies previous ones. For example, plant families have been reclassified based on the evidence collected from cladistics.

Classification is the process by which living organisms are sorted out into different groups or taxa based on their similarities.

Taxa (singular: **taxon**) are the groups, such as phyla, genus and species, into which taxonomists sort the organisms they are classifying, so taxonomy is the science of classification of living things.

A **binomial nomenclature** or **binomial system** is a naming system where two names are used: in the Linnaean binomial system the names used are the genus name and the species name.

Nomenclature describes a system for assigning names to things.

6.7 Classification

There are many different living organisms in the world. Sorting out living organisms based on similarities is called *classification*. Grouping organisms makes it easier for scientists to study their characteristics.

Carl Linnaeus (figure 19), was a Swedish botanist. His *binomial nomenclature* is the basis for the system (often known as the Linnaean system) currently used by zoologists, botanists and taxonomists all over the world. Key features of Linnaeus's system of classification are:

- the placement of organisms into groups (*taxa*) based on their shared characteristics

- using scientific names to identify each organism instead of the myriad of names that different communities may have for it. For instance, Linnaeus used a genus and species name for each organism he described and classified.

DP ready | **Theory of knowledge**

As part of his work in classifying living organisms, Carl Linnaeus (also known as Carl von Linné or Carolus Linnaeus) classified humans into four groups based on both physical and social traits. While in the 18th century this may have been acceptable, current standards view his descriptions as racist. How did the social context of Linnaeus's scientific work affect the findings of his research? Do we need to consider the social context when evaluating ethical aspects of knowledge?

Figure 19. Carl Linnaeus (1707–1778)

Figure 20. The quokka, native to Western Australia and its surrounding islands, is known by the scientific name *Setonix brachyurus*

Hierarchy of taxa

Living organisms are arranged in hierarchies of groups (taxa) based on their shared features. In the hierarchy of taxa, groups of living organisms are placed in large groups which are further divided into smaller groups. All organisms are classified into three domains: Archaea, Eubacteria and Eukaryota. Eukaryotes are further classified into seven main principal taxa: kingdom, phylum, class, order, family, genus and species (figure 21). Organisms can exist in only one group at each level of classification. For example, an organism can only belong to one kingdom.

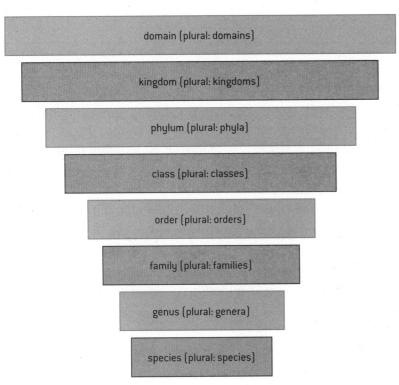

Figure 21. Hierarchy of taxa

DP ready   Approaches to learning

Using mnemonics may help you when studying. To memorize the hierarchy of taxa, you may use the mnemonic in table 6.

Table 6. Mnemonic for hierarchy of taxa

Domain	**D**ear
Kingdom	**K**ing
Phylum	**P**hilip
Class	**C**ame
Order	**O**ver
Family	**F**or
Genus	**G**ood
Species	**S**oup

Table 7 shows the seven principal taxa for the classification of humans.

Table 7. Classification of humans (*Homo sapiens*)

Taxa	Human
Kingdom	Animalia
Phylum	Chordata
Class	Mammalia
Order	Primates
Family	Hominidae
Genus	*Homo*
Species	*sapiens*

Binomial system of nomenclature

In the binomial system, every organism is given a name using Latin words. Linnaeus developed the binomial system so that the same name is used around the world. The name given to living things is made of two parts: genus and species.

For example, the scientific name for humans is *Homo sapiens*. "Homo" is the name of the genus while "sapiens" is the name of the species.

When writing the scientific name of any organism, the following rules must be followed:

- If the name is printed, italics are used (for the genus and species names). If hand-written, it is underlined.
- The genus name is written first. The first letter is always capitalized. For example, *Homo*.
- The species name is written second. The first letter is never capitalized. For example, *sapiens*.
- Both names are joined together: *Homo sapiens*.
- You can use abbreviations to write the scientific name. The genus name can be abbreviated to just its initial: *H. sapiens*.

Table 8 shows some examples of the binomial names for different species.

Table 8. Examples of the binomial system of names for some species

	Genus	Species	Complete name	Abbreviation
Lion	*Panthera*	*leo*	*Panthera leo*	*P. leo*
Garden pea	*Pisum*	*sativum*	*Pisum sativum*	*P. sativum*
Wolf	*Canis*	*lupus*	*Canis lupus*	*C. lupus*

> **DP ready** **Nature of science**
>
> Scientists cooperate and collaborate with each other. For example, the binomial system of names for species is universal. Instead of using local names for species, scientists use the binomial system to identify a species.

The three domains

The first classification system was based on classifying living organisms into five kingdoms: Prokaryotes, Protista, Plantae, Fungi and Animalia. However, when the nucleic acids of prokaryotes were compared, prokaryotes were split further into two different groups: Archaea and Eubacteria. Therefore a higher grade of taxonomic group was needed to reflect this, now called a domain. Differences in nucleic acids and proteins are used to reflect evolutionary relationships. As a result, living organisms are classified into three domains:

1. **Eubacteria**: "true" bacteria; prokaryotes without a nucleus and without membrane-bound organelles.

2. **Archaea**: "ancient" bacteria; are also prokaryotes. Most groups live in extreme environments. These include:
   a) Methanogens, which live in anerobic conditions such as swamps and the gut of animals. They use carbon dioxide to make methane.
   b) Thermophiles (heat lovers), which live in very hot habitats such as hot springs in volcanic areas.
   c) Halophiles (salt lovers), which live in habitats with a very high salt content such as the Dead Sea.

3. **Eukaryota** (also known as Eukarya): single-celled and multicellular organisms which all have their DNA contained in a nucleus. This domain includes the kingdoms of plants, animals, protists and fungi.

DP ready | Theory of knowledge

In 1977, Carl Woese discovered that the evolutionary line descent of archaean bacteria is separate from the line of evolutionary descent of "true" bacteria. Therefore, he divided prokaryotes into two main groups: Archaea and Eubacteria. Many scientists objected to his new classification. To what extent is conservatism in science desirable?

Figure 22. Many archaea are extremophiles; that is, they are often found in extreme environments. For example, archaea of the genus *Picrophilus* are able to survive in the extremely acidic hot springs of Hokkaido, Japan

The animal kingdom

The multicellular organisms in the animal kingdom can be further divided into two major groupings, animals without backbones (invertebrates) and animals with backbones (vertebrates). Figure 23 shows examples of invertebrates and figure 24 shows examples of vertebrates.

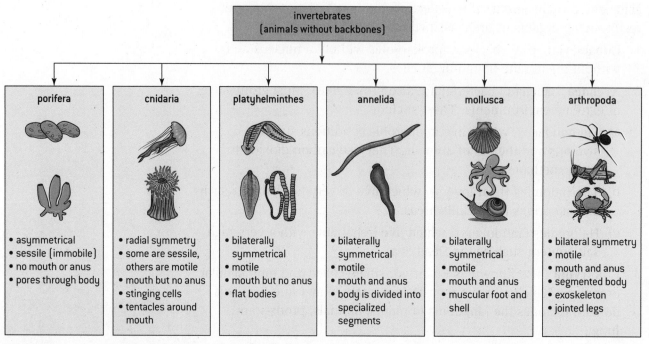

Figure 23. Invertebrates (animals without backbones)

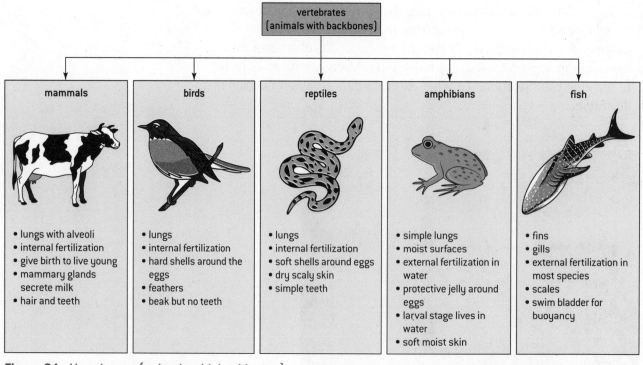

Figure 24. Vertebrates (animals with backbones)

The plant kingdom

Some of the major groupings within the plant kingdom can be seen in figure 25.

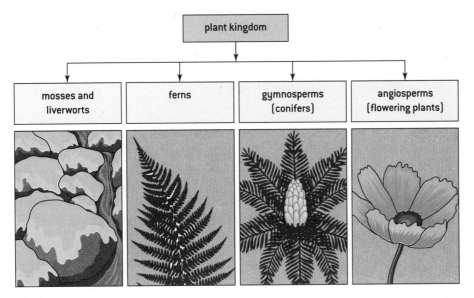

Figure 25. The plant kingdom

Dichotomous keys

The *dichotomous key* is a tool that allows you to determine the features of a living organism. The term "dichotomous" means "divided into two". This is because the dichotomous key always gives you two choices in each step. You will always start at the first stage and then answer the question to move to the following stage.

 Key term

A **dichotomous key** is a tool used to identify living organisms.

Worked example: Using dichotomous keys

5. Use the dichotomous key below to identify the organism in figure 26.

Figure 26. What is this organism?

1. Has legs............................ Go to 2
    Has no legs .... Go to 3

2. Has 6 legs .... Insect
    Has 8 legs .... Spider

3. Has a shell .... Mollusc
    Has no shell ....Go to 4

4. Has a segmented body .... Earthworm
    Has no segmented body .... Tapeworm

Solution

The living organism in the picture has no legs, so go to 3. Since it has a shell, it is a mollusc.

Question

7 Use the dichotomous key in the worked example to identify the
   organism in figure 27.

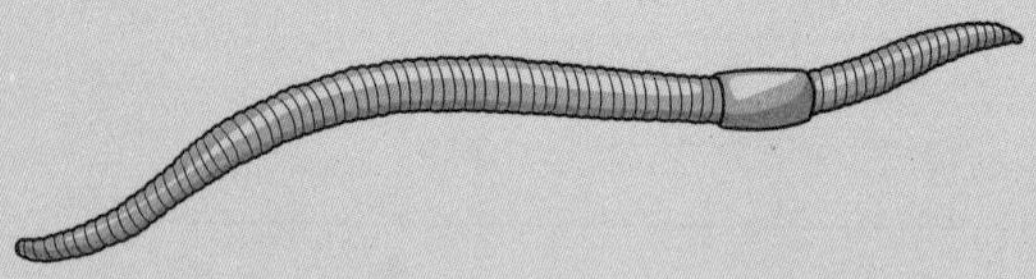

Figure 27. Clue: This is not a chicken

Chapter summary

In this chapter, you have learned about:

- [ ] An ecosystem is a community of living organisms interacting with the non-living things in their environment. It consists of biotic and abiotic factors.

- [ ] Living organisms can be divided into two main categories based on the way they obtain their food: autotrophs and heterotrophs.

- [ ] Depending on the method of food intake, heterotrophs may be consumers, parasites, detritivores and saprotrophs.

- [ ] Consumers could be herbivores, carnivores or omnivores.

- [ ] Some organisms use a combination of different modes of nutrition and can act as both autotrophs and heterotrophs depending on the surrounding conditions. These organisms are known as mixotrophs.

- [ ] Food chains show the movement of energy from one organism to the next.

- [ ] Producers, primary consumers, secondary consumers and tertiary consumers are examples of trophic levels in food chains.

- [ ] A food web is a diagram that shows the relationships between organisms in a community. The arrows in the food web show the direction of the energy flow.

- [ ] Amount of energy flow in a food chain is represented numerically by a pyramid of energy.

- [ ] Ecosystems require a continuous supply of energy to replace energy lost as heat, whereas nutrients are recycled.

- [ ] The greenhouse effect is a natural phenomenon caused by the greenhouse gases.

- [ ] The theory of evolution and natural selection discusses how species change over geological time.

- [ ] Antibiotic resistance in bacteria is an example of evolution in response to an environmental change (in this case, the use of antibiotics).

- [ ] Speciation occurs when populations become so different that they can no longer interbreed.

- [ ] Cladograms are tree diagrams that show a group of organisms (clades) that have evolved from a common ancestor. Cladograms are used to deduce evolutionary relationships.

- [ ] Nomenclature describes a system for assigning names to things.

- [ ] The binomial system uses two names for each thing it describes: the genus and the species.

- [ ] Classification is the process by which living organisms are sorted out into different groups based on their similarities.

- [ ] Living organisms are classified into three domains: Eubacteria, Archaea and Eukaryota (Eukarya).

Additional questions

1. Distinguish between autotrophs, heterotrophs and mixotrophs.
2. What is main role of saprotrophs?
3. Discuss reasons why the bars of a pyramid of energy differ in size.
4. What are the main sources of carbon dioxide on Earth?
5. Outline the evidence for evolution provided by homologous structures.
6. What are the causes of variation in a species?
7. Explain how homologous anatomical structures are considered evidence of evolution.
8. Explain the use of cladograms.
9. Explain how natural selection can lead to evolution.
10. Distinguish between Cnidaria and Mollusca.

7.1 Approaches to your learning

You should concentrate on developing a good all-round understanding of biology. While knowledge and rote learning are important, a good understanding of the key concepts is crucial. All areas of biology share common ideas such as energy, cells and chemical and physical processes, so make a habit of trying to link the topic you are studying now with what you learned earlier. The *Internal link* icon (⊗) in this book shows areas where learning can be joined up.

Our brains are good at forming links—make the most of this skill. The Diploma Programme is all about learning and developing concepts across all your subjects. Aim to become an effective student biologist with good learning skills and an independent approach to your work. In the Programme you should seek to develop the following.

- *Communication skills*
  All scientists need to communicate their findings and results effectively. These important skills will pervade your entire Diploma work. As a Diploma student you will also transfer these skills to your other group subjects, the Theory of Knowledge and the Extended Essay.

- *Self-management skills*
  You need to know the right conditions for learning for yourself, and to be able to create them. There is advice on this later in this chapter.

- *Research skills*
  These will help you not just in your IA project but in collecting ideas and material for the Theory of Knowledge essay and the Extended Essay.

- *Thinking skills*
  You should develop the ability to think critically, to innovate and be creative within a scientific context.

- *Social skills*
  These are your skills at being an effective member of a group, whether in the group 4 project, in a collaboration in group 6, in a fieldwork survey in group 3 or just organizing an activity with friends.

These skills make up the five areas addressed in the IB by the phrase *Approaches to learning* (◉). Throughout your course (and beyond) ask yourself the following questions:

- What are my present skills in biology; how well are they developed?
- Are any of my skills too weak for comfort; how can I improve them?
- What additional new skills do I need; how can I learn them?

7.2 Good study habits

The start of a new course—whether moving to the IB Diploma from a GCSE, national qualification or Middle Years Programme background—is an appropriate time to review your study habits. A good approach to learning is vital if you are going to make the most of the programme. In the spirit of the five *Approaches to learning* skills above, review how you study and ask yourself what improvements you can make.

Help your brain

The key to good study is to integrate all the information you are learning and to retain it. You need long-term, not short-term, memory, and the way to transfer from short- to long-term is by reviewing material regularly.

Our memory of a lesson (or any other experience) is at its best immediately thereafter; then the memory starts to fade. To retain the memory longer you need to review the material about one day later. But that memory will not be permanent either. The next time for a review is after a week, and then once again after a few weeks. Continue this review process and eventually the memory will become unshakeable. Think about lessons when you were younger: the teacher set homework on the day, gave a test on the material perhaps a week later, and then had a major test on material at the end of a term or semester. This was designed to make you review the work over and over again.

Some of the advice on note-taking later in this chapter relates to this idea of review.

Planning for study

You need to plan your work schedule if you are to review regularly as suggested above. The Diploma Programme is demanding of time. You take six subjects and Theory of Knowledge classes together with other school commitments and your creativity, activity and service (CAS) projects. Then there is your social and family life outside school. You need to use a planning diary into which you put *all* your commitments. Later, it will be important to include the revision you intend to undertake. There is advice about this later.

When and where to study

Everyone is different. Some people work best in the morning; others peak in the evening. Play to your strengths here; don't force a work pattern on yourself that your brain and body says is wrong. If you are a morning person, get up early and do your work before school.

Include breaks in your planned schedule – remember how the brain operates. Everyone needs 10–15 minutes to settle down to effective study, so don't plan to do any important work then. Equally, you need breaks of a few minutes at regular intervals, perhaps every 20–40 minutes. But do not take too long over the break as then you will need settling time again. Making and drinking a cup of coffee is about the right length for the break; a long phone call or a protracted social media session is not.

Effective note-taking

At this stage in your education, you should be learning to take your own notes. The ability to make clear notes is a vital skill in any job or university. The start of the Diploma course is a good time to take stock of your note-taking talents.

- You need to choose whether a hand-written or electronic system is for you. It's also possible to write by hand directly onto a tablet using a stylus. Whichever approach you use, make sure that the notes are permanent and secure. Paper notes should be filed carefully in a sensible order. It is not necessary to carry your entire Diploma notes around in a very large folder. Some students do this and are then unfortunate enough to lose their folder.

- With electronic notes it is important to have a regular backup routine to store the files. This storage should be secure and preferably cloud-based. However, if this is not possible, store your files in separate places in several locations. As with notes on paper, this is information that you cannot afford to lose.

- Some people scan handwritten notes and store the files electronically. In this case, choose an unambiguous and systematic set of file names.

- If you intend to keep notes using a computer or tablet then there are several solutions: programs such as Microsoft OneNote and Google Keep are designed for notes. Once again, choose whatever works for you.

Whether you choose a handwritten or computer-based system, there are several ways to construct lesson notes and the other documents you produce for your schoolwork. When you return to the notes you must be sure that you will still understand them. Figure 1 shows an example of Cornell note-taking:

	Keywords:	Notes: Membrane transport
	simple	• simple diffusion
	diffusion	- particles move across membrane
		- E is not needed
		- passive
		- e.g. CO_2 moves from cells $\rightarrow$ blood
	facilitated	• facilitated diffusion
	diffusion	- channels needed
		- E is not needed/passive
		- particles (polar) move across
	summary:	

simple diff	facilit. diff
• in both particles move across membrane	
• in both E is not needed/passive	
no channels	channels needed

Figure 1. Cornell note-taking

- Use visual aids to help you remember things such as mind maps (figure 2).

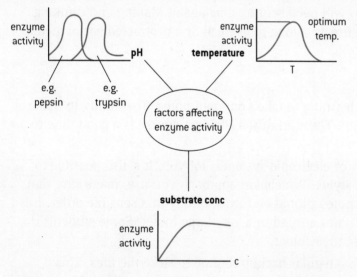

Figure 2. Factors affecting enzymes activity summarized as a mind map

7.3 Academic honesty

Throughout your Diploma Programme, not just in biology, you will be collecting the words and thoughts of others, from the opinions of classmates through to the writings of distinguished scientists. It is entirely appropriate that you should quote other people in support of your work. What is not appropriate is passing these words off as your own to gain credit. This is academic dishonesty and, if you are quoting verbatim from others without crediting them, is plagiarism. Refer back to the Frederick Gowland Hopkins quote at the start of Chapter 1—this is an example of a quote with a detailed reference.

Making full and consistent notes allows you to note down references for later use.

7.4 Preparation for examination

Attempt as many past papers as you can to learn about the structure of the paper and the style of questions. In the Introduction of this book, there is a table that illustrates the structure of each paper in the DP Biology HL and SL examination. Make sure to understand the structure of each paper and the type of questions expected in each section.

Effective revision

Effective revision is essential to focus on the main concepts that you need to focus on for the exam. The following tips may help you improve your revision techniques:

- Find a quiet place for your revision.
- Start revision as early as possible.
- Use your notes and add to them if needed.
- Practise past papers, with an initial focus on questions that cover concepts you struggle with.

Understanding examination questions

Internal link

The complete list of DP Biology command terms is present on page 166 in the appendix of this book.

It is essential to understand the language of DP questions before you attempt to answer them. To do so, you need to be familiar with the command terms for DP Biology. These are verbs that indicate the depth of treatment required in the questions. Command terms are usually found at the beginning of the question. Notice the increase in difficulty as you move from one assessment objective to the other.

Here is an example from paper 2 Biology SL:

> **Compare and contrast** how pyruvate is used in human cells when oxygen is available and when oxygen is not available.
>
> **Outline** the process of glycolysis.
>
> **Explain** the process of aerobic cell respiration after glycolysis has occurred.

The command terms are in bold. These terms inform you of how to answer a question. For example:

- The term 'compare and contrast' means to give reasoned similarities and differences between two ideas or concepts.
- The term 'outline' means to give a brief account or a summary.
- The term 'explain' means to give a detailed account that explores all areas of the argument.

Here are few tips to follow with regard to command terms and when answering questions:

- If the question is to **compare and contrast**, you might find it useful to put your answer in a table. Remember that you need to give similarities and differences. For similarities, use the term 'both'. When it comes to differences, remember to give the parallel attribute for the second idea or concept.

Idea 1..	Idea 2...
• Both 1 and 2 are ...... • Both 1 and 2 have	
1st difference	1st difference (parallel)

- If the question is to **distinguish**, then you just need to give differences, also in a table.
- With questions such as **calculate**, where you need to give a number as the answer, remember to show your working and give units with your answer.
- If the question is to **discuss**, divide your answer into: arguments for and arguments against.
- If the question is to **evaluate**, divide your answer into strengths and limitations.

- If the question is to **draw**, make sure of the following:
  → Your diagram or graph is at least half a page
  → Use a dark pencil when drawing
  → For labelling, draw a line and not an arrow
  → Use a ruler for straight lines
  → Make sure you are precisely labelling the part needed.

7.5 Tips when approaching examination papers

Biology is a challenging subject as it requires memorization of a large amount of information. Practising past papers will help you feel comfortable and will enable you to explore and be familiar with typical DP exam questions. There are some tips that might help you maximize your mark when approaching exam papers.

Paper 1 (multiple choice questions)

- Read the questions very carefully and highlight key words. Remember that you can use a highlighter during the examination, so make use of it.
- Pay attention to statements that include 'not' or 'except'.
- Read all the options before choosing any.
- Eliminate the options which are obviously wrong to reduce the number of options.
- If you don't know the answer to a question, circle it and move to the next. You can come back to it later.
- Review your answer sheet and make sure that you have answered all questions.

Paper 2 and 3

- Both papers are mixtures of structured questions, data analysis and extended response questions.
- Make sure that you are familiar with the command terms for IB Biology as previously discussed.
- Circle the command term to make sure you understand what is asked.
- Pay attention to the marks awarded for each question and organise your time spent on each question accordingly.

Data-analysis questions

- Such questions involve a graph or a table that requires interpretation.
- Always read the introduction very carefully. Highlight important information.
- Study the graph/table very carefully. Pay close attention to the axis and axis labels.
- There is always a pattern in the questions – they will start with *calculate/state/identify*, then *compare and contrast/distinguish*, and lastly *hypothesize/discuss*. The difficulty of the questions increases from one question to another.
- If the question is to *identify/state*, you can refer to the graph/table to get the answer. Remember to add the unit.
- When the question is to *calculate*, a common question is to calculate the percentage difference. In this case, use the following formula:

$$\% \text{difference} = \frac{\text{New value} - \text{Old value}}{\text{Old value}} \times 100$$

- When the question is to *describe* a relationship/ trend, remember to:
  → Look for positive or negative correlations
  → Describe it:
    "as x increases…y increases/decreases"
  → Ask yourself whether it is a weak or strong correlation

→ Describe the increase/decrease as significant/slight/little

→ Comment on how steep the line is

→ For a bar graph, look at the highest and lowest points

→ When the x-axis is time, ensure you refer to it in the description: "over time, the metabolic rate increases"

→ Look if there is a change in the trend: "becomes constant/levels off/reaches plateau"

→ Look for fluctuations in the data.

- When the question is to *distinguish*, mention the differences and remember to use comparative terminology

- When the question is to *discuss* or *evaluate*, note the following:

  → Highlight the idea to be evaluated or discussed

  → You need to refer to all previous graphs/tables to answer the question.

  → Make sure you read the introduction before each graph as it may have important information.

  → Divide your answers into:

Arguments for:	• Refer to every graph and try to support the idea from each graph/table • Refer to the introduction/ information written before each graph/table

Arguments against:	Comment on validity of data—for example, on standard deviation/ error bars
	   • Large standard deviation/error bar—data is widely spread around the mean/ high variation/ high uncertainty/ invalid data
	   • Small standard deviation/error bar—data is close to the mean/ low variation/ low uncertainty/ valid data
	   • In a bar graph, if error bars overlap—no significant difference
	   • In a bar graph, If error bars don't overlap—there is a significant difference

	Range
	   • Large range—high variation—high uncertainty—invalid data
	   • Small range—low variation—low uncertainty—valid data

	Sample size (number of trials/ participants/ data)
	   • Large sample size—more reliable—valid data
	   • Small sample size—not very reliable—invalid data

	Limitations in the sample selected
	For example, focusing on females not males/on animals not humans
	Other factors may have an effect such as genetic factors, environmental factors and the duration of time for conducting the experiment.

Extended response questions

- Read the whole question before you choose the one you want to answer.

- Choose the question you feel more confident to answer: where you can attempt to answer all parts of the question (a, b and c).

- Read the question carefully and highlight the key terms.

- Develop a simple mind map to organize your ideas before answering the question.

- Use bullet points to answer these questions to organise each point logically.

- Include all relevant ideas; don't eliminate ideas that you may think are common sense.

- If you provide a graph in your answer, make sure that you label the x- and y-axis.

- Don't select the drawing question unless you can provide a clear drawing with all labels.

- Remember that there is one extra mark for the quality of construction of your answer, so make the structure of your answer as clear as possible.

Mathematical skills

Mean

The *mean*, also known as the *average*, can be calculated by dividing the sum of all values in a data set by the *sample size* (the number of values obtained). When conducting an experiment, you may repeat a specific measurement several times. Calculating the mean of the values obtained from these measurements gives a more *precise* value.

$$\text{Mean} = \frac{\text{sum of all samples}}{\text{sample size}} = \frac{\sum x}{n}$$

Range

The *range* is calculated by finding the difference between the largest and the smallest measured values in a data set. We usually calculate the range to learn about the spread of data.

Range = largest value – smallest value

Mode

The *mode* is the value in a data set that occurs most often.

Median

The *median* is the central value of a data set. You can determine the median by listing all the values in ascending order and finding the value directly in the middle of the list.

> **Worked example: Calculating the mean and range**
>
> A student did an experiment measuring the height of five bean plants over one week. Data collected were as below. Find the mean and range of his data.
>
> | Height of bean plants (cm) | 7.0 | 6.0 | 9.0 | 6.0 | 10.0 |
> |---|---|---|---|---|---|
>
> *Solution*
>
> $$\text{Mean} = \frac{\sum x}{n} = \frac{(7.0 + 6.0 + 9.0 + 6.0 + 10.0)}{5} = 7.6 \text{ cm}$$
>
> Range = 10.0 – 6.0 = 4.0 cm
>
> Mode = 6.0 cm, as this value occurs the most frequently.
>
> To the find the median, list the values in ascending order:
> 6.0, 6.0, 7.0, 9.0, 10.0
>
> Median = 7.0 cm, as this is the central value in the list.

Percentage change

In experiments, we are often interested in how a variable changes under a given set of conditions. A useful value to quantify this change is the *percentage change*. To calculate the percentage change, find the difference between the two values you are comparing (the original value and the new value) then divide by the original value. Finally, multiply the answer by 100%. If the answer is positive, then the value has increased. If the answer is negative, then the value has decreased.

$$\text{Percentage change} = \frac{(\text{new value} - \text{original value})}{\text{original value}} \times 100\%$$

Worked example: Calculating the percentage change

The mass of a potato cylinder was measured before and after it was submerged in saline solution. Results were obtained as below. Calculate the percentage change in mass for the potato cylinder during the submergence period.

Mass of potato cylinder in saline solution (g)	
Initial mass (before submergence)	Final mass (after submergence)
2.2	1.9

Solution

Percentage change $= \dfrac{(1.9 - 2.2)}{2.2} \times 100\% = -14\%$

Therefore, there is a 14% decrease in the mass of the potato cylinder.

Standard deviation

The *standard deviation* is a measure of how spread out your data are. It uses the mean of your data set as a benchmark – the variation of each data point compared to this mean value will determine the value of the standard deviation. This is calculated using the following formula:

$$\sigma = \sqrt{\dfrac{\sum (x - \bar{x})^2}{n}}$$

where σ = the standard deviation, $\sum$ = the sum of, x = each value in the data set, $\bar{x}$ = mean of all values in the data set and n = the number of values in the data set.

- If you obtain a large standard deviation, this indicates that your data are widely spread around the mean which shows high variation. This means that your data are less precise and has high uncertainty, and therefore less valid.

- If you obtain a small standard deviation, this indicates that your data are close to the mean which shows low variation. This means that your data are more precise and has low uncertainty, and therefore more valid.

Worked example: Calculating the standard deviation

A student did an experiment counting the number of oxygen gas bubbles produced by two different plants in 10 minutes. The student carried out this experiment six times for each plant. The results are indicated in the table below. Calculate the mean and standard deviation for each of the data sets recorded for each plant.

Experiment number	Number of oxygen gas bubbles in plant A	Number of oxygen gas bubbles in plant B
1	4	7
2	3	1
3	5	2
4	3	4
5	4	2
6	5	8

Solution

Plant A mean $= \dfrac{(4+3+5+3+4+5)}{6} = 4$ bubbles

Plant B mean $= \dfrac{(7+1+2+4+2+8)}{6} = 4$ bubbles

For plant A, SD $=$

$$\sqrt{\dfrac{(4-4)^2+(3-4)^2+(5-4)^2+(3-4)^2+(4-4)^2+(5-4)^2}{6}} = 0.812 \approx 1$$

For plant B, SD $=$

$$\sqrt{\dfrac{(7-4)^2+(1-4)^2+(2-4)^2+(4-4)^2+(2-4)^2+(8-4)^2}{6}} = 2.646 \approx 3$$

Internal link

In this worked example, one significant figure is used for values of the mean and the standard deviation. Refer to the Maths skills box on page 7 in **1 Cells** for an explanation of significant figures.

Experiment number	Number of oxygen gas bubbles in plant A	Number of oxygen gas bubbles in plant B
1	4	7
2	3	1
3	5	2
4	3	4
5	4	2
6	5	8
Mean	4	4
SD	1	3

We can note that though the mean is the same for the two sets of data, the SD for plant B is larger than that of plant A. This indicates that the data collected for plant B are widely spread out around the mean, which indicates a relatively high variation in the data. This shows the data collected for plant B are less precise and therefore less valid.

Error bars

Error bars can be used to give a visual representation of the standard deviation of data on a graph. Looking at an error bar for a given mean value on a graph will you give you an idea of how spread out the data are with respect to this value, and therefore how valid the data are.

For example, if the standard deviation is 1 cm, this means that the error bar will be drawn a unit of 1 cm above the mean and 1 cm below the mean. These bars are then capped at either end (figure 1).

- ***The smaller the error bar,*** the smaller the standard deviation. This indicates that your data are close to the mean which shows low variation. This means that your data are more valid, more precise and have low uncertainty.

- ***The larger the error bar,*** the larger the standard deviation. This indicates that your data are widely spread around the mean which shows high variation. This means that your data are less valid, less precise and have high uncertainty.

Two examples of error bars used in a bar graph and a line graph are shown below:

a)

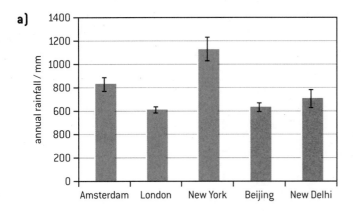

b)

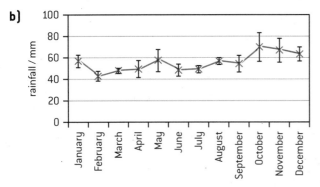

Figure 1. Graphs with standard deviation error bars

Charts and graphs

When collecting data, we usually organise it in a table, and then use a graph or a chart to present it. There are several types of charts and graph. You need to select the one that is the best representation of your data. Some of most common types of graphs and charts are discussed here:

- *Bar charts* are used to compare between groups or categories. For example, when comparing the mass of leaves of a plant growing in the shade and light (figure 2).

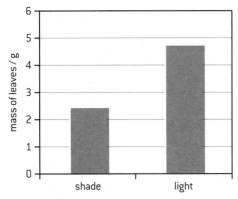

Figure 2. An example of a bar chart

- *Histograms* are used to show the frequency of occurrences for a specific category. For example, the category could be the length of leaves of a plant growing in the shade, with the histogram showing the number of leaves for each different leaf length (figure 3).

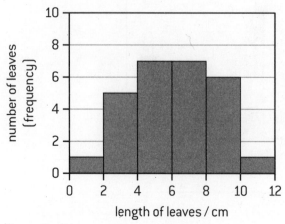

Figure 3. An example of a histogram

- *Line graphs* are used to show trends, most commonly for how a variable might change over time. Each data point is connected to the following data point by a straight line. For example, the change in average global temperature over the last 100 years (figure 4).

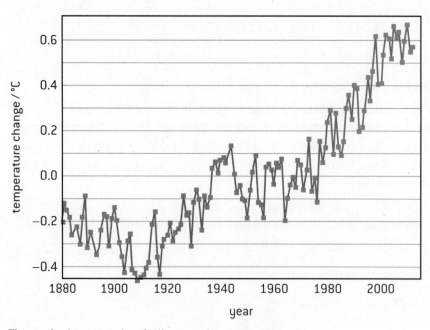

Figure 4. An example of a line graph

- *Scattergraphs* are used when looking for a relationship (or correlation) between two variables. If a relationship between the variables exists, you can plot a line of best fit, or trendline, to describe this trend. For example, the effect of increasing enzyme concentration on the rate of a reaction (figure 5).

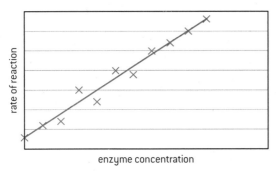

Figure 5. An example of a scattergraph

- *Pie charts* are used to compare the percentage of categories as part of a whole. Adding all of the percentages will give 100%. For example, the percentages of different living organisms in a certain area (figure 6).

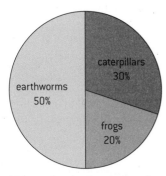

Figure 6. An example of a pie chart

ICT skills

Data loggers

You can use *data loggers* when collecting large data over a period of time in an accurate and efficient way.

For example, you can use an oxygen probe to measure the concentration of oxygen gas produced in photosynthesis over a period of time.

Simulations

Advances in technology allow us to replicate real-world biological processes using computer *simulations*. The benefit of these simulations is that they open up our experimental freedom as we are able to change parameters and conditions of an experiment easily. They are also useful in aiding our visual understanding of processes in biology that we are unable to observe first-hand. You may experience the use of simulation at DP Biology in two ways:

- As a demonstration by your teacher to reinforce a concept
- As part of your practical or internal assessment.

You can find plenty of free simulations available on the internet. You need to be careful when selecting a simulation for the internal assessment—make sure the simulation is accurate and relevant to the concept you are studying.

Software for data processing and analysis

When data are collected, you can use specific software to present your data in a table and then plot a graph accordingly. For example, data loggers can directly process the collected data into a graph. Microsoft Excel is another example of software that could be used for presenting and processing data. It also contains specific functions for analysing such data. For example, you can use Excel to easily calculate the mean or standard deviation.

Glossary of command terms

As discussed in *7.4 Preparation for examination*, the following command terms are used in examination questions to indicate the depth of treatment required. It is essential for you to be familiar with these key terms and phrases.

Assessment objective 1	
Define	Give the precise meaning of a word, phrase, concept or physical quantity.
Draw	Represent by means of a labelled, accurate diagram or graph, using a pencil. A ruler (straight edge) should be used for straight lines. Diagrams should be drawn to scale. Graphs should have points correctly plotted (if appropriate) and joined in a straight line or smooth curve.
Label	Add labels to a diagram.
List	Give a sequence of brief answers with no explanation.
Measure	Obtain a value for a quantity.
State	Give a specific name, value or other brief answer without explanation or calculation.

Assessment objective 2	
Annotate	Add brief notes to a diagram or graph.
Calculate	Obtain a numerical answer showing the relevant stages in the working (unless instructed not to do so).
Describe	Give a detailed account.
Distinguish	Make clear the differences between two or more concepts or items.
Estimate	Obtain an approximate value.
Identify	Provide an answer from a number of possibilities.
Outline	Give a brief account or summary.

Assessment objective 3	
Analyse	Break down in order to bring out the essential elements or structure.
Comment	Give a judgment based on a given statement or result of a calculation.
Compare	Give an account of the similarities between two (or more) items or situations, referring to both (all) of them throughout.
Compare and contrast	Give an account of similarities and differences between two (or more) items or situations, referring to both (all) of them throughout.
Construct	Display information in a diagrammatic or logical form.
Deduce	Reach a conclusion from the information given.
Design	Produce a plan, simulation or model.
Determine	Obtain the only possible answer.
Discuss	Offer a considered and balanced review that includes a range of arguments, factors or hypotheses. Opinions or conclusions should be presented clearly and supported by appropriate evidence.
Evaluate	Make an appraisal by weighing up the strengths and limitations.
Explain	Give a detailed account including reasons or causes.
Predict	Give an expected result.
Sketch	Represent by means of a diagram or graph (labelled as appropriate). The sketch should give a general idea of the required shape or relationship, and should include relevant features.
Suggest	Propose a solution, hypothesis or other possible answer.

Index

Key terms are in **bold**.